CHRISTIANITY

AND

ECOLOGY

Edited by

Elizabeth Breuilly

and

Martin Palmer

CASSELL

Cassell Publishers Limited
Villiers House, 41/47 Strand, London WC2N 5JE, England
387 Park Avenue South, New York, NY 10016–8810, USA

© World Wide Fund for Nature 1992

First published 1992

British Library Cataloguing-in-Publication Data
A catalogue record for this book is available from the British
Library.

Library of Congress Cataloging-in-Publication Data
Available from the Library of Congress.

ISBN 0–304–32374–8

Cover picture: *The Third Day: The creation of the land and the
vegetation*, from the banner paintings by Thetis Blacker in
Winchester Cathedral, used with permission
Panda symbol © 1986 World Wide Fund for Nature

Typeset by Fakenham Photosetting Limited, Fakenham, Norfolk
Printed and bound in Great Britain by Mackays of Chatham plc,
Chatham, Kent

CONTENTS

ACKNOWLEDGEMENTS

The following material is used by permission:

pp. 2, 55, cartoons: Fran Orford;

p. 22, illustration by Horace Knowles: Copyright © The British and Foreign Bible Society, Swindon, England, 1954, 1967, 1972;

p. 66, illustration: The British Library;

p. 80, illustration: The Master and Fellows of Corpus Christi College, Cambridge;

pp. 98–9: Janet Morley, from *Till All Creation Sings*, Christian Aid Autumn Anthology, 1989;

pp. 106–7: from a Confession of Faith from a Service for Human Rights from Chile, *Confessing Our Faith Around the World*, IV: *South America* (p. 65), WCC Publications, World Council of Churches, PO Box 2100, 1200 Geneva 2, Switzerland, 1985;

pp. 99, 107, 110, drawings by Caro Inglis from the booklet for the Creation Festival Liturgy, Coventry Cathedral, 9 October 1988; p. 109, illustration;

pp. 109–10, extracts from the Creation Harvest Liturgy, Winchester Cathedral, 4 October 1987: World Wide Fund for Nature.

INTRODUCTION

As recently as 1985, it would have been almost impossible to find a book on Christianity and ecology. Most ecologists thought Christianity was more of a problem than a friend, and few Christians seemed to be aware of the importance of the messages the environmentalists were giving the world. Since then much has changed. World-wide, the Churches, in different ways and at different speeds, have been responding to the challenge: 'What does Christianity have to offer to the environmental debate?'. This book seeks to capture something of that debate and to offer models and examples of Christian environmental concern.

The book falls into four main sections:

A—an introduction to the crisis and to Christianity's role within it;

B—an exploration of some of the roots of Christian thinking about creation, the environment, and our place within it;

C—accounts of some of the many historical strands and ways in which this has been worked out in practice;

D—a wide-ranging look at how Christians are responding today.

In the first section, Dr Freda Rajotte, until 1991 a member of the World Council of Churches' Church and Society unit, with Elizabeth Breuilly, a consultant for ICOREC, draws upon her experience world-wide to present the stark picture of what we have done, and shows some of the attitudes that have contributed to this—attitudes which have flourished in largely Christian cultures. This is not a comfortable section. It cannot be, for the sheer scale of the threats that face us means that we have to rethink both our faith and our attitude to nature.

The search for the roots of our behaviour leads us to section B. Dr Ruth Page, associate Dean in the Faculty of Divinity at Edinburgh, examines in a sharp and critical way the images of nature and of our role in nature within the Bible. She raises considerable questions about the extent to which the Bible can actually supply us with workable models for a new relationship with nature, given the history of Christian abuse of nature.

From the biblical tradition, the debate moves to the traditional teachings and understandings of the Church down the centuries. Here, Metropolitan John of Pergamon, a senior Orthodox bishop and theologian, explores the way in which the Church has defined humanity and its relationship to nature through many centuries and against many different philosophies. His exploration of the Orthodox understanding of humanity and nature brings a vision, which is often missing in Western Christianity, of both our power and our humility within nature.

In section C, we move from the theological and philosophical to the practical. What has the Church in the past done about nature? One of the earliest models of environmental Christianity is the Benedictine movement. It has been claimed that it was the Benedictines who saved Northern Europe from what some see as the first major environmental crisis brought about by humans— the collapse of the Roman Empire. Sister Joan Chittister explores the practice of her tradition and gives a practical example from her own community on the shores of Lake Erie, in the USA.

St Francis is undoubtedly the best-known of Christian saints who have been concerned with ecology. In his article, Father Peter Hooper, director of the Franciscan Study Centre in Canterbury, UK, looks at why Francis's teachings were so radical and at the failure of Franciscans until very recently, to follow the teachings of their founder.

With the coming of the Protestant movement in the early sixteenth century, Christianity underwent the most extensive transformation in its history. What did this do to our relationship with nature? In his article, Martin Palmer, director of International Consultancy on Religion, Education and Culture (ICOREC) and an Anglican theologian, looks at the very dramatic and devastating changes which certain aspects of the Protestant revolution unleashed upon the world. In particular the impact of Calvinist thinking is shown to have been a major contributor to the growth of an exploitative attitude towards nature.

Finally, in section D, Freda Rajotte looks at the variety of responses and reasons for those responses within the Church today. She paints a tough, demanding but ultimately hopeful picture of the future of our planet.

The book is designed to offer thought and reflection—which is why the editors have inserted questions to help you reflect in most of the articles. It can be read straight through, like a journey through the faith. Or you may find it is more helpful to wander within it, listening to the different voices and experiences which have come together to create this book—voices and experiences which give some hint of the vast and complex picture of Christianity today and its relationship to the natural world.

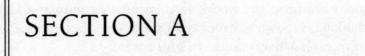

SECTION A

1 WHAT IS THE CRISIS?

Freda Rajotte
with Elizabeth Breuilly

THE ILLNESS AND ITS SYMPTOMS

More and more people throughout the world are worried about the way the natural world is being harmed and destroyed. They see how the sheer numbers of human beings are causing severe damage to the world. The earth's biological systems such as food chains and habitats, and its physical systems such as the water cycle and the ozone layer, are being harmed by what we do. All the different problems: drought, famine, global warming, the spread of deserts, vanishing forests, pollution of the seas, the extinction of species, can be seen as signs of the same disease—a disease which may yet prove fatal. It looks as though the earth's ecology is dying, and its life-support systems are closing down. It seems that we are locked into a spiralling pattern of self-destruction.

What is this disease affecting the earth? What caused it, and how can it be cured?

Christian leaders and thinkers have joined the voices sounding the alarm. Pope John Paul II summed up the dismay felt by many: 'We begin to ask how we can have destroyed so much. We also ask anxiously, if it is possible to remedy the damage which has already been done.' Patriarch Dimitrios of the Orthodox Church also sounds the warning: 'We cannot continue plundering God's creation without reaping the results of its eventual destruction.'

What sort of creatures are we, that we have done this? What sort of creatures must we become in order to halt and reverse what is happening?

1

WARNING:
HUMANITY CAN SERIOUSLY DAMAGE YOUR HEALTH

Fran Orford

Environmental destruction is not only a danger to us all, it is a sin against God. Pope John Paul II underlines this: 'I wish to repeat that the ecological crisis is a moral issue.'

But the Church's response to the environmental crisis has been slow and hesitant. Edward Echlin shows how in the late 1960s and early 1970s, Church leaders were already speaking about pollution, conservation, and expanding consumption of natural resources. He goes on:

> Had these utterances been taken up they would be recognized today as prophetic. But they were not taken up. Leadership in restoring solidarity with the wider living community passed to the United Nations, WWF, National Trust, Council for the Preservation of Rural England, Yorkshire Wildlife Trust, the generous young people of Greenpeace and Friends of the Earth—and half a hundred other similar groupings.
>
> (*The New Road*, bulletin of the WWF Network on Conservation and Religion, no. 4)

Why is this? It may be because the Church has spent many centuries encouraging some of the ways of thinking which have led us into this mess. So before we can see how to respond, and how Christians all over the world are responding, we must look at what has been wrong, not only with the world, but with the Church. For the answer is not simply a matter of teaching the people of the world to care for their environment. If planet earth is sick, then the Church is part of the illness. Jesus said: 'Why do you observe the splinter in your brother's eye and never notice the plank in your own? How dare you say to your brother, "Let me take the splinter out of your eye", when all the time there is a plank in your own? . . . Take the plank out of your own eye first, and then you will see clearly enough to take the splinter out of your brother's eye' (Matthew 7:4–5). It will take a repentant, restored and revitalized Church to respond adequately to the present crisis. In Jesus, the Christian Church holds the secret of the remedy for the earth's sickness, but we must heed the challenge: 'Physician, heal yourself!'

Questions for discussion

1 (a) How long ago did you first become aware that there is an 'ecological crisis' (or whatever phrase you want to use)?
  (b) How did you first become aware of it? (e.g. from newspaper, television, school, church etc.)
  (c) How often are you aware of Christians discussing this problem, whether at local, national or international level?

2 Is it fair to ask: 'What sort of creatures are we, that we have done this?' How would you begin to answer this question?

THE CHRISTIAN FAITH—HAS IT ANYTHING TO OFFER?

Can the Christian faith offer any insight into these problems? Many people are asking how the churches can respond to the ecological crisis. Of course, more and more individuals and groups are becoming involved in local efforts to conserve or re-store the environment, as part of their Christian concern for the world.

But the most important task of religious communities is to provide a vision to sustain and support environmental action:

— How do we see the world?
— What is our relationship with our fellow human beings and with the rest of creation?
— What is God's will for the world?

The answers to these questions, and the power with which they live in our minds and hearts, will make all the difference to the way we relate to other people. This might be with friends and family, or in the many other ways that we relate to each other: as employer and employed, as buyer and seller, as teacher and taught, as helper and helped. These relationships exist not only between individuals, but also between whole societies, nations and continents. But they are all shaped by what we believe about the world, as individuals, as a society, as a Church.

How much do our beliefs matter when we are faced with the possibility of the complete collapse of the earth's ecology? Does it matter whether we believe in the various 'isms' that are put forward for our consideration—eco-feminism, utilitarianism, bio-centrism, and so forth? Clearly, when we are faced with an im-mediate emergency, such as cleaning up an oil spill, it does not matter. But in the long term, if we are to continue to live on this earth, human society needs to work on a totally different system from that which dominates the world today. We need to give things a different value, a value which does not rest on money or power. We need to be able to work for the welfare of the planet, not because we know we ought to, not because we are afraid of dying if we don't, but because we have seen something of great value that is worth all our best efforts.

This sort of deep change needs a clear vision—a vision of the world as God created it and intended it to be, a vision of the way that human beings should relate to each other and to the world. Our efforts to change in small ways will be lost and confused and overwhelmed by the vastness of the problem, unless we have seen and believed in the vision of a new order where everything is looked at in a different way.

So theology is vitally important. By 'theology' we don't mean a dry, academic subject. Theology is the way each person understands their faith, what their experience tells them about God, what they believe God says about their experience and about their place in the world. The theology that can provide ways to approach the ecological crisis is likely to be very different from traditional theology as it is discussed and taught in seminaries, cathedrals and conference halls. Interpretations of the Bible and of the Christian faith must be based on the insights and struggles of people suffering under oppressive systems, for Jesus made it clear that they are of the first importance in the Kingdom of Heaven—and to speak without experience is to produce only empty words.

Christianity can only be true to itself when it is part of the hearts and lives of its people, in whatever situations those people find themselves. Today the Bible is read and interpreted by Christians all over the world: by the unemployed, women, the poor, the dispossessed, the homeless, the hungry, those in despair—by the 'wretched of the earth'. Their experience and their understanding of the Christian faith are the new theology that we must learn—a theology based on working together; a theology based on the stresses and strains of practical action; a more humble and less arrogant theology; a revolutionary theology. Examples of some of these new ways are given in the final chapter of this book.

The new voices which are increasingly being heard from outside traditional Church structures are not only asking us to look at new understandings and new ideas: they are showing more clearly what has been wrong for centuries, with those structures and Church thinking. The environmental crisis arises largely from the way that Western society thinks and works. Western society has developed from the structures and traditions of the Christian Church in the West.

We saw on p. 1 that we as human beings have to face what kind of creatures we are, to have created such environmental damage and injustice. This is painful enough. But as well as this, the Church has the pain of recognizing its own need to change. The 'bride of Christ', the 'body of Christ on earth', the Church has been involved in hundreds of years of injustice—has allowed colonialism, slavery, the subjection of women and the conquest and ruin of nature.

What then are the faults and mistakes which the Church needs to acknowledge and repent of if it is to offer any worthwhile treatment for the earth's sickness which it has helped to bring about?

Questions for discussion

1 In what ways does the Church (local, national or international) make you angry?

2 Does the Christian faith as it is usually presented have relevance to your own experience:
   — all your experience?
   — some of your experience?
   — none of your experience?

3 What questions should the Church speak about if it is to be more relevant to the issues that you think are important?

THE CAUSES OF THE DISEASE

Many of the environmental problems of the world arise from the culture that has spread from Europe and North America over the last 500 years, affecting almost the whole world. And this culture is based on Western versions of Christianity, and has in turn affected the way that Western Christians think. It is not that Christianity necessarily thinks this way, or that there are no Churches with a different view. But the powerful Churches, the Churches whose voices have been heard, have been European in origin. Let us look at some of the attitudes which have helped to lead the Church and the world down the road to destruction.

(a) The centre of the universe?

For centuries, we in the West have thought of ourselves, of humanity, as the most important thing in the universe. To some extent this is only natural. Everyone is self-centred to some extent—just look at a baby or a toddler. As we grow up we gradually learn to consider the needs and feelings of others. But we still do not look beyond our own circle. And for us in the West, 'our own circle' revolves around human beings and planet earth as our province. It is not so in every culture: Australian Aborigines, for example, quite naturally see certain other creatures as part of their own family, and the land and its plants and water as being closely related to them.

It is Western Christianity that has taught us to see ourselves as the centre of things with the right, even the duty, to conquer, subdue and have dominion over nature. Early scientific thought reinforced this view of ourselves, by seeing the earth as the centre of the universe, with the planets, the stars and the sun all circling round it.

When Copernicus showed in the sixteenth century that the earth is not the centre of the system that moves planets and stars through space, this rocked not only the scientific world, but the religious world as well. If the earth was not the centre of the universe, then it began to look as if 'man' was not as important as he had first thought. And that was not all. Human self-importance was threatened again when Darwin showed that, far from being totally separate from the rest of creation, human beings have slowly evolved from it, and are part of it.

But somehow, while most Christians have taken on board the scientific discoveries, they have pushed aside what they mean in terms of belief, attitude and lifestyle. We read the words of Psalm 8, and hear only half of them. We are quite happy to assert that:

> You have made [man] little less than a god,
> You have crowned him with glory and splendour,
> made him lord over the world of your hands,
> set all things under his feet. . . .

But we have forgotten the awe and humility expressed in the first half of the psalm:

> I look up at your heavens, made by your fingers,
> at the moon and stars you set in place—
> ah, what is man that you should spare a thought for him,
> the son of man that you should care for him?

Yes, human beings are important—but not through any merit or power of their own. If God has given humanity a high place it is not because of what we are, but simply because he chose to. God's decision came first, and our importance follows from it, not vice versa.

Towards the end of the twentieth century there are new experiences and new knowledge that may at last help to ram the message home and make us more humble. For the first time we have seen pictures of our planet from outer space, and these have helped us to take on board our own smallness in the universe. Science and technology are discovering patterns in the universe which suggest that we may not be as unique a life-form as we think we are. It is beginning to look as though humanity is just a small part of a much larger living body referred to as the 'biomass' or 'biosphere'—the whole earth and the life on it.

Just as in a human body you cannot separate the life of the lungs from the life of the stomach or the life of the brain, it is beginning to look as though you cannot separate the life of human beings from the life of the trees, the insects or the seas.

Until very recently we still spoke and lived as if the earth, the plants, the animals, all the rest of creation, were there only for our benefit. Any land not cultivated, any rivers not harnessed for power generation, were regarded as useless or worthless: the value of anything was measured solely in terms of its usefulness to 'man'. Because we thought that a world 'out there' existed which was somehow separate from ourselves, and not related to us, we felt free to exploit, dominate and subject it.

Today there are voices raised which blame much of today's ecological crisis on Western Christianity which has so blindly put humanity at the centre of the universe. Today many theologians are considering once again the human place within the biosphere.

Questions for discussion

1 Christians pray about the natural world—to give thanks for it, to ask for good weather, etc. Choose one aspect of the natural world and think what *it* might want to pray to God about *you*.

2 What is your reaction to Psalm 8? Do you identify most with the humility of verses 3-4 or the power expressed in verses 5-6? (Be honest!)

(b) Dismembering the world

Even before the human race began to tear apart the earth in a physical sense, we had dismembered the earth in a spiritual sense in our way of thinking about it and studying it. For we have tended to think of the environment as something 'out there' for us to possess, to buy and sell, to make a profit from, or to study. We forget that we ourselves are members of it, and cannot exist without it.

This was not always so. Some of the very early Christian writers referred to the world as a living being, as holy, as a mystery of which we are a part. In the third century AD, one of Christianity's earliest great theologians, Origen, spoke of the world as 'an immense living creature which is united by one soul, namely the power and reason of God', in whom everybody and everything exists.

To despoil the earth is to desecrate something that is holy. To cause the suffering and death of part of creation is to cause the suffering of God, for it is in God that all things have their being. Many people, for example Celtic Christians, Native American peoples and Siberian tribes such as the Tungka, think of the earth as a great mother who brings forth and nurtures all life. To pollute the earth is like spitting on one's mother. One Canadian Indian theologian referred to the whole environment as 'all of my family'.

Taking this argument a step further, it may be that humans are not alone in having a spiritual sense. Spirituality is an aspect of the mystery of life that humans share with all creation. It is not something alien or strange that has been parachuted in from outside, but something which has evolved gradually as living beings became aware of themselves. The Old Testament writers certainly saw the whole created order as being aware of its maker, and bound in eternal praise to him:

All things the Lord has made, bless the Lord:
give glory and eternal praise to him. . . .
Sun and moon! bless the Lord:
give glory and eternal praise to him.
Stars of heaven! bless the Lord:
give glory and eternal praise to him.
Showers and dews! bless the Lord:
give glory and eternal praise to him. . . .
Every thing that grows on the earth! bless the Lord. . . .
Seas and rivers! . . . Sea beasts and everything that lives in the water! . . .
Birds of heaven!. . . Animals wild and tame!. . .
Servants of the Lord! bless the Lord:
give glory and eternal praise to him.

(From the Song of the Three Young Men, in the version of the book of Daniel
found in the Greek Old Testament and Roman Catholic editions of the
Bible.)

If we see 'man' as a spiritual being who just happens to be set down in a world which is nothing but 'things', we are tearing apart both ourselves and the world.

As scientific and technical knowledge grew, and as it became more and more separated from religious knowledge, people began to see the world as made up of 'things'. Nature was seen as a gigantic machine, made up of separate parts or atoms that were interchangeable, rather like simple building blocks. The universe was reduced to bits of lifeless matter that obeyed a series of laws. All objects from the sun to a mustard seed responded to the general laws of physics, chemistry and biology, laws such as gravity, growth or respiration. And increasingly the Church followed this view.

This view of reality produced a vast amount of detailed information, because people could study the various parts without being distracted by the complex whole. We owe much valuable knowledge to this way of studying, but it had several unfortunate consequences:

— it reduced God to being only the 'giant watchmaker', who had set the world in operation according to unchanging rules. God was no longer seen as being involved in the operation of the world and necessary to its existence day by day;

— the relationships between all things were ignored;
— scientific research became separated from ethics and morality: it was as if God had simply handed us the 'giant watch'
   and said 'Here's a puzzle for you—there's a prize if you can
   find out how it works!';
— the 'prize' came to be seen only in material terms, in the
   ability to manipulate the world to our own advantage.
   There was no other meaning or purpose to existence.

As scientists and technologists discovered more and more about
the 'laws' of nature and how to manipulate them, people began to
think that human beings could go in a different direction from
nature. For centuries, people were fooled into thinking they could
master and redesign nature. Many phrases in our language point
to this: land is 'developed' when it is built upon; diseases are
'conquered'; rivers are 'controlled'; even the genes that differen-
tiate different living things can be 'engineered'. Human beings
invented new substances never seen before—non-biodegradable
substances, poisonous substances. On and on they went, invent-
ing ever more and newer, in meaningless numbers. When living
beings are affected by uncontrolled and meaningless growth we
call the disease 'cancer', and we look to see what has caused it—
the carcinogen. If we think of the earth as suffering from a kind of
cancer, we do not have far to look for the carcinogen. The more
numerous people are, the more goods and services they 'need', the
more space they occupy, and the more resources and consumer
goods they demand—and the more waste and pollutants are
produced.

Until recently, people believed without question that technology and development would eventually be able to produce all
the goods and all the food needed by the world's rising popu-
lations. We in the rich world felt confident that eventually the
whole world would be able to enjoy cars, refrigerators, televisions
and electronic gadgets at the same level as ourselves, and that as
each new piece of technology was invented, it would make life
easier all round.

It is only recently that it has become clear that whatever the
earth gives us in fuel, minerals or food, is a limited supply. Tech-

nology has simply allowed us to use up that supply even more quickly. The earth's resources cannot cope, cannot renew themselves, and will be destroyed if this goes on. Technology can never provide the cure, because, as we have said, the problem is not just a practical one, but arises from people's attitudes, beliefs and morality. If technology is seen as the cure, it can only make the disease worse.

A similar problem arises with the idea of 'development'. For many years, development was seen as the solution to the problem of poorer countries, and many projects were started with high hopes. But in many cases these 'developments' actually made the problem worse. There are two main reasons for this.

One is that economic and industrial development gives more power to those who are already rich and powerful: to large land-owners, governments and multinational industries. Development projects often take land away from small farmers who grow food, and use it instead for products which are sold abroad.

The other reason lies in the very idea of development. If we think only in terms of economic growth, we are bound to see both people and the land as things to be used to help us achieve it. That way of thinking takes away the value that people, and land, and all created things, have of themselves. People are not simply workers: each one is an individual with a unique value. No other person could ever be a 'substitute' for me, even if she did the same work in the same way. And the land is not simply wealth locked up, ready to be taken—it is our home, the basis for all life.

There will be no solution to poverty and exploitation, of either the land or the people, unless we give up the idea that economic growth is the aim, and the land and people simply the tools or the means to achieve that aim. This way of seeing things gives no value and no purpose to people or to the world. It is yet another way in which we are tempted to 'dismember the earth'.

Questions for discussion

1 Scientists seek to find out more about the physical world and how it works. They may do so for many different reasons. List some possible reasons that you think are good, and some that you think are bad.

2 Are there some questions about the physical world that science will never be able to answer?

3 Are there some questions about the world that science should not try to answer?

4 Think of an example of development in your own area—a new road, shopping centre, community centre etc.
   (a) Who made the decision for the development?
   (b) For whose benefit is it intended?
   (c) Who has actually benefited?
   (d) Who has suffered?
   (e) Do you think, overall, it has been a success or a failure? Why?

5 In what senses can we think of ourselves as being separate, or not separate, from the earth?

(c) Other divisions

We have seen how this dividing of ourselves from nature has been fostered by the Church and has led to problems. But it is not the only division that causes problems. It seems to be a universal law of human nature and human sinfulness that people divide themselves from each other. The Church needs to ask itself: 'What is our response to divisions of belief, divisions of lifestyle, divisions of culture?' There are so many different branches and traditions of the Christian Church that there will be many and conflicting answers. Here we look at some of the problems and some of the principles.

(i) Us and Them—other faiths

In the past, Europeans have generally assumed, rather arrogantly, that Christian thinking provided the only true source of knowledge about the world, and gave a universal set of moral codes for behaviour. This attitude justified not only mission activity, but the conquest of non-Christian sociieties, the Crusades, and other 'holy wars'.

Sharing one small planet and faced by global threats, people are now becoming aware of the need to speak to each other, and genuinely to try to hear what the other is saying. This can only happen in an atmosphere of trust and respect. We have to try to understand the other point of view, not just in our own terms, not just saying, 'What does this mean to me?' or 'Are there any ideas here that I can use?'. We have to ask instead, 'What does this mean to you?', 'What would you like me to see and understand?'. We have to take on board the 'otherness' of the other. This is not easy, but trying to hear each other in this way enriches both sides.

Dialogue between faiths should allow each to draw from the others the resources to help face today's threats to the world. Several steps have been taken in this direction. The Roman Catholic Church and the World Council of Churches have each established units for dialogue with other faiths. In 1986 the WWF (World Wide Fund for Nature) called upon religious leaders from different faiths to meet in Assisi, Italy, for a celebration and declaration of co-operation between religion and the conservation movement. Each faith brought its own insights, and each faith spoke to its own people, but together the New Alliance of Conservation and Religion was founded.

Dialogue between religions is not only for the leaders. Individuals must learn and are learning to work together, and as they do so are learning more of each other's beliefs and traditions.

(ii) Denominations

The Church proclaims oneness, unity and wholeness. But the reality has been and still is very different. The gulf between different forms of Christianity often seems greater than that between Christians and non-Christians who are active in any given field such as famine relief or conservation.

The Churches must recognize their need to come together and the sinfulness of the way they have rejected each other in the past. All denominations have spiritual gifts to share, whether monastic traditions, social action, art treasures, evangelism, music or ritual. It requires the gifts and insights, the work and commitment of all, to strengthen each other in facing and working out the problems of our times.

(iii) Modern and indigenous, colonizer and colonized

Over the last 500 years, since Columbus opened up contact between Europeans and the rest of the world, European culture,

How Western Christians saw themselves in relation to others. From Rev.
J. G. Wood, *The Natural History of Man*, 1870.

European economic methods, and European Christianity have
expanded together. And with them went the ideas about development, about the world as an object to be used, which we have
looked at earlier.

In colonial situations the results were often disastrous for the indigenous peoples. The fact that they had often lived in a state of balance with the land for many hundreds of years was ignored. Their lands were occupied and divided up, and large areas were devoted to export crops. In some places, taxes were imposed by the colonial authorities with the sole purpose of forcing indigenous people to work for cash rather than their traditional subsistence farming.

This sort of 'development' provided immediate and vast cash profits for the colonizers. It transformed land into resources and people into labour. Everything was given a cash value, and the whole earth became a planet for the taking. Islands, seas, forests and prairies were stripped and ravaged. Sandalwood, whales, seals, sea-elephants, buffalo were plundered. Indigenous peoples were evicted from their lands or eliminated by genocide. Any person or thing that could not be exploited or sold was regarded as valueless. A holocaust of destruction was unleashed.

For nearly 500 years Churches stood approving or at least silent as this colonization went on. A few Churches or individuals spoke out against the cruellest excesses, but until the nineteenth century even slavery was accepted as normal. Few Church voices were raised against the mass slaughter of animals or the destruction of forests and soils. Today, many Churches work with and support the struggles of indigenous people to live sustainably on their own land.

Young, independent Churches in Africa and Asia are now interpreting Christianity in their own contexts, for example as Nigerians or as Chinese, instead of taking on board European culture along with Christianity. Sometimes they bring insights to biblical understanding that are based on their traditional ideas and beliefs about human community, and of the sacredness and unity of nature. It is hard for those of us who have lived with a long tradition of European Christianity to recognize that much of what we hold dear is more European than specifically Christian—but that is what must be done.

(iv) Women
There is no doubt that the Church has been guilty of suppressing,

ignoring and marginalizing women, as have many societies all over the world. Throughout the Third World, and in many industrial nations, women outnumber men amongst the very poor, the oppressed and the powerless, as well as in church congregations. Women are rarely at the centre of things.

But however much one agrees that, for the sake of justice, women's voices should be heard more and women's contribution recognized, what has this to do with the environmental crisis? Modern economics tends to ignore the work of nurturing, feeding and sustaining life, and because this work does not earn money, labels it as 'unproductive'. This attitude degrades both women and nature. Recognizing nature's importance in sustaining life is the same as recognizing women's importance in sustaining the family: maybe neither of them produces wealth in money terms, but both are vital.

It has often been noticed that where men and women do not cooperate equally and with respect for each other, that is where poverty and underdevelopment are most found. There is a very close connection between the status of women and the status of the land. The lower the status of women, the higher the birthrate, infant mortality, and the strain on natural resources.

In Africa, for example, the soil in many places is exploited for gain, and becoming steadily poorer and poorer. Exactly the same is true of women, and development agencies and the Churches have realized that no one can tackle the problems of the land without tackling the problems of women. In many parts of the world, women are taking the lead in steps to halt deforestation, such as the Chipko movement in India, where women hug trees to save them from felling, and the Green Belt movement in Kenya which initiates tree planting projects. Women's tasks include providing wood for fuel, water, and nearly 80 per cent of the food crops that the people live on. It is women who are the main farmers; women who care for the land. Their actions, experience and opinions are therefore vital.

The Churches have played a large part in keeping women down and silencing them. They must begin to play a part in raising them up and listening to them, within Church power structures as well as in the outside world.

17

Questions for discussion

1 Take each of the divisions discussed in this section. On which side of each one does your situation place you? Do you find it hard to understand what the other side's experiences are? How might you come to understand better?

2 Are division and disagreement a necessary part of our human nature? How should we try to cope with them?

A long hard look

The Church, especially where it has been in positions of power in powerful countries, has often been corrupted by that power. Comfortable and self-centred ways of thinking have crept in, and the Church has found itself little by little separated from the sufferings of the world. The Christian Church needs to take a long, hard look at itself, at what it tells the world in word and deed. We cannot hope to do right until we have seen where we have been wrong. We also need to see what traditions and teachings have come down to us to help us see a way forward.

SECTION B

2 | THE BIBLE AND THE NATURAL WORLD

Dr Ruth Page

THE OLD TESTAMENT

> O Lord, how manifold are thy works,
> In wisdom thou hast made them all.
>
> (Psalm 104:24)

This verse sums up what the Bible has to say about the whole various world of heaven and earth as its writers knew it. Some modern thinkers may suppose that the world is the result of chance, but the writers of the Bible are clear that it is the result of God's wise 'making'. In biblical times however, no one would have thought of the world as a random happening. People of all nations believed in some god or gods and some act of creation. The Old Testament had no need to assert that the world came about through a creative purpose rather than by chance—everyone agreed on that. What the Old Testament does emphasize is that the world is the direct work of one transcendent God who rules its past, present and future, in contrast to the view that the world was the home and manifestation of many gods. To Israel's neighbours the Canaanites, a storm, for example, was the manifestation of the storm god Baal, but in the Bible nothing in heaven or on earth has divinity within itself. A storm in the Old Testament is not the appearance of God but rather his tool or instrument: 'who makest the clouds thy chariot, who ridest on the wings of the wind, who makest the winds thy messengers, fire and flame thy ministers' (Psalm 104:3f.).

The Bible never refers to the natural world as an independent being or set of beings. Even the abstract term 'creation' does not appear in the Old Testament and is rarely used in the New. Wherever the natural world is mentioned in the Bible the writers are concerned with the creatures God has made, each with its own relationship to God the creator. This relationship comes about through divine action, for God does not create a world and then abandon what is made. The writers of the Bible hold that *everything*—the components, inhabitants, location or weather of the world—is a creature deriving from God and dependent on God. It is all under God's control, God's care and, when necessary, God's judgement.

Psalm 104, from which I have already quoted, is a good example of this belief. The Psalmist exults in the variety and order of creation with its mountains, rivers, birds, trees, wild goats, lions and humans. It is through God's ordering that 'the beasts of the forests' like the lions come out at night but return to their dens by sunrise, the time when 'man goes forth to his work and to his labour until the evening' (v. 23). Everything has its place in the wisdom of God's ordering: mountains are settled where they are, and when God's finger touches them they smoke; seas, which are perilous, chaotic things liable to flood the dry land, are fixed in their place by divine decision. God's work is plain to anyone who will look. In Psalm 19 'the heavens are telling the glory of God and the firmament proclaims his handiwork'. The firmament was understood like a bowl inverted over the earth containing the heavens. To the Psalmist the firmament was like visible speech—a language which expressed the greatness of God who created its vastness and set orderly lights within it: the sun to shine by day and the moon and stars to shine by night. The Old Testament shows humanity as part of this orderliness, set on earth to tend and preserve it, yet with the capacity to go astray and disrupt the whole interrelationship.

Israel was a small country surrounded by larger, stronger ones. How did it arrive at so total a vision of creation by one all-powerful God? The nations around Israel described their gods and told their creation stories from the sense of a world full of divine presences, and the experience of powers and forces in nature as men and women struggled for existence. People had to bow to the

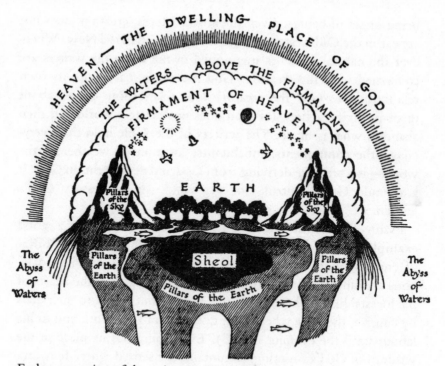

Early conception of the universe. Horace Knowles.

forces of nature which controlled their lives, and these were seen as the forces of the gods: they were not objects but personalities much greater than humans. In that ancient world no sharp distinction was made between the natural and the supernatural, so the bull, for instance, could be both a natural bull and the representation or presence of 'the bull of heaven', as the gods Anu and Enlil in Mesopotamia were called.

Creation stories within such beliefs often involved the struggle between two opposing powers or gods. Babylonian stories, for instance, involved Apsu and Tiamat, gods of the original watery chaos. They have intercourse and give birth to other gods. One of these, Marduk, defeats Tiamat in battle, and by tearing her body apart creates heaven and earth. Only Israel (and, in a different way, Greek philosophers) moved away from this kind of creation story, and it is important to realize how different Israel's belief was from the countries around it.

The difference with Israel is that belief in one almighty God did not arise from experience of the powers of nature. Israel's belief was mainly shaped by the experience of the Exodus, when their God had freed them from slavery in Egypt and had led them through the wilderness to the promised land. That meant that the Israelites were built into a nation by God, who had said to them 'I will take you as my own people, and I will be your God' (Exodus 6:7). All other belief was based on that promise and on the story, told and retold, of how God had brought them out of Egypt. No other gods were to be worshipped but their God who saved them, and God's power was total: God was stronger than the Pharaoh who had tried to keep them in slavery, stronger than the Red Sea which stood between them and freedom, and strong enough to cause water to flow in the desert. So when Israel settled down as a nation and began to wonder how the world was made, that could only be understood as the act of this single, all-powerful God.

The Old Testament is, first and foremost, a Jewish book. It was written by Jews from their special experience of God. Christian belief is based on Jewish belief, and the picture presented in the Old Testament of one God who created everything that exists, is a fundamental of Christian belief as well.

Questions for discussion

1 How would you sum up the difference between the Israelites' idea of their one God, and other ancient peoples' idea of their gods?

2 If Psalm 104 had been written by someone who believed in the world as the home of many gods, how would it have been different? Take one or two verses and rewrite them from this point of view.

The book of Genesis, the first book of the Bible, contains two different accounts of how the world was created and how the relationship between human beings and God developed. Like many books in the Bible, Genesis was not written by one author, but collected together from many different sources, some written, some handed down by word of mouth. So it is not surprising to find two different accounts side by side.

23

The first, in Genesis chapter 1, was put together by one or more Jewish priests writing in the sixth or fifth centuries BCE (Before Common Era) and is called the Priestly document. It is a stately, ordered account of God's creative action day by day over six days. In it God, who existed before everything, effortlessly creates all things by word alone: 'God said "Let there be light", and there was light. . . . God said, "Let the waters teem with living creatures and let birds fly above the earth within the vault of heaven" and so it was'—and so on. For this writer God does not have to come down to shape creation: God's word is sufficient and effective in bringing all things into being.

The differences from the Babylonian tales described on p. 22 are striking: there is no duality of sexes here, no procreation or divine birth, no struggle between opposing powers. Instead the eternal God, transcendent and alone, creates everything out of nothing. There is even a special verb in Hebrew, *bara*, used only of God's creating, to show how unique and divine that act is. Therefore all the creatures that are brought into being are distinct from God and are not to be confused with the one who made them, but belong to God. This account says of each part of creation as it is made that 'God saw that it was good': that is, it is fitting for its purpose and fits into the total design. Later on, especially in Christian thinking, some people began to reject the physical, material world as if it were evil, and to favour the spiritual as something purer. There is nothing of that here. But matter is not praised or glorified either— it is meant to be 'fitting', fulfilling its intended function.

The story in Genesis 1 is an account of the power of God's word which commands a movement from chaos to order, from un-created darkness to created light in which order can be seen, from formless water to the formation of land, seas and the heavens. The earth is created to be mother to all plants and animals. In the same way the seas give birth to their living creatures. Productiveness is a characteristic of creation. Plants are not said to be living crea-tures because they appeared to lack what the writers of the Old Testament take to be the basis of life, 'breath'. The Hebrew *nephesh*, breath, is what sea creatures, land animals and humans share to make them alive. There is no distinction among them on this score.

Although humans and animals both have *nephesh*, there is a clear difference between them according to this account, in that man and woman are made in the image of God. The Hebrew word used for image (*tselem*) is that also used in the Old Testament for the appearance of an idol or statue. This implies that the whole person including physical appearance was intended here by the use of the word. Later theology tended to insist that it was humanity's intelligence or reason which showed it to be the image of God. (This idea is discussed in more detail in Metropolitan John's chapter on p. 51.) But the emphasis of Genesis 1 appears to be more concerned with the idea that humanity was created in order to represent God in action on earth. Intelligence is necessary for this but there is more to it than that. One writer discussing this takes Genesis 1:26 and explains it by adding words to it: 'Then God said, Let us make man (to act) as our representative (on earth), (to be) someone (enough) like ourselves (to be able to understand what we were about in creating the world)' (J. C. L. Gibson, *Genesis Vol. 1*, Daily Study Bible, Edinburgh, St Andrew's Press, 1981, p. 74).

The point of humanity as the image of God is that we were created for a special purpose. God has dominion and power *over* the earth, and human beings were created to represent that power *on* earth. The words used for this dominion and power are certainly fierce, with the sense of trampling or stamping on the rest of creation. Other chapters in this book refer to the accusation that Christian thinking has encouraged domination rather than dominion over the world's resources (see pp. 7ff., 86ff.). The strong words used in Genesis, chapter 1 seem at first sight to confirm this accusation, but only if they are taken out of context. For if we look at the whole story, humanity is the representative of God who has just brought about a pleasant, orderly world in which creatures can flourish and multiply! The creation of man and woman is followed by the sabbath when God rested since 'all the host' of creation was in place and could begin to live together the life it was created for, populating the earth in the process. At that point chaos has been transformed thoroughly, happily and fruitfully into an ordered universe.

A second, very different account of creation follows immediately in Genesis 2. The people who put Genesis together did not try to decide whether one was 'right' and the other 'wrong', nor did they try to iron out the differences between them. They let both stand alongside each other, possibly so that both might express in their different ways beliefs about God's action and purposes in creation. The second account is some centuries older than the first and is told in the form of a story rather than the step-by-step description of God's actions in Genesis 1. In Genesis 1 the transformation is from chaos to an ordered world. In Genesis 2 it is from desert to garden: 'There was as yet no wild bush on the earth nor had any wild plant yet sprung up, for the Lord God had not sent rain on the earth, nor was there any man to till the soil' (Genesis 2:5). God is described as planting a garden in Eden, and then creating a man and a woman to live there and look after it. It is important that Eden was a garden because it shows that the Old Testament ideal is for cultivated land being productive. With the dry and stony soil of Palestine it is understandable that writers would feel no appreciation of wild, uncultivated country and thought rather in terms of well-watered, fruitful farms.

What happens within the garden is told as a story with vivid detail and compelling plot. It is told to show why things are as they are now. Here God is not the distant and effortless creator that we see in Genesis 1, but acts like a potter making things out of clay. The name 'Adam'—the man that God moulded out of clay—comes from the Hebrew *adamah*, ground. Once more God's gift of breath is the sign of life. Adam is given the duty of working in the garden preserving it from damage and encouraging it to fruitfulness (Genesis 2:15). God, who wants the best for creation, is concerned at Adam's solitude. Animals and birds are created to be companions to him, though he finds their companionship limited. Then God provides him with a woman from his own body. Because they are like each other, but different, they can be true companions to each other. Variety makes for harmony.

The harmony and blessedness continue as long as everything and everyone keeps to the role that God laid down for them. God had commanded the man and the woman not to eat the fruit of one particular tree; they could eat anything else that grew in the garden, but not that one. But they cannot resist the temptation to step over the boundaries God has given them. They disobey God and they suffer divine judgement. Their punishment is to be turned out of the garden into a harsh and difficult world where farming is hard work and child-bearing is painful—in other words, into the kind of world the writer knew. This is how he explains all the difficulties in the way of human or earthly fruitfulness: they arise from humanity's desperate desire to be like gods.

In the story, the woman is the first to be tempted to eat the fruit, and she offers it to the man. Some people have used this aspect of the story to argue that therefore women are more disobedient to God than men, and therefore more to blame than men for what is now wrong in the world. But the story does not read that way. The man and the woman both eat the fruit, and both make excuses when their crime is found out.

Questions for discussion

1 Read the two accounts of creation in Genesis 1 and Genesis 2. How do you react to each of them? If you were writing a poem or a piece of music about creation, which account would you mainly draw on? If you were trying to explain the Christian view of the world as it is now, which account would you draw on?

2 What does the account in Genesis 2 say about what people are like today, and why they behave as they do?

3 When you read the story of God creating the universe, what images come to mind? Draw a picture entitled 'Creation' and discuss it with one or two others.

The rest of the Old Testament is mainly concerned with the relationship between God and Israel—the creation of a people rather than the creation of the world. We saw earlier (p. 23) that Israel's whole idea of God was based on what God had done for them and the solemn agreement between them known as the covenant. It involved a promise from God, but the people also had to play their part, by keeping God's laws. Within that covenant was included human care for useful plants and animals, which were to be respected. Just as, within the covenant, one must live in fairness with other humans, so one must deal fairly with farm animals and the land under cultivation. 'You shall not oppress a hired servant who is poor and needy' (Deuteronomy 24:14), but equally 'You shall not muzzle an ox when it treads out the grain' (Deuteronomy 25:4). Clearly it makes good sense to keep both human and animal servants alive and strong enough to work, but the Bible goes beyond this, for it is concerned with their well-being. The covenant makes a relationship between human beings and God, and within this it is clearly right to respect what God has made, even while one makes use of it.

The same respect is shown in the instructions concerning the sabbath—every seventh day, which is set aside for rest and worship. It is not just human workers who are to rest, but the land and animals as well (Exodus 23:10–12). Similarly every seventh year was 'a sabbath of solemn rest for the land' (Leviticus 25:4), when all lay fallow. Finally, after seven times seven years came the year of jubilee, the year of liberation within Israel in which the farmlands and vineyards also had their share: 'You shall neither sow, nor reap what grows of itself, nor gather the grapes from the undressed vines' (Leviticus 25:11). 'The land', God reminds them, 'is mine' (Leviticus 25:23). The poor were allowed to gather what they could from the fields during the sabbatical year. So the sabbath combines honour to God, concern for the poor, and a sensitivity to the need of the land for rest after years of production.

In the New Testament Jesus insists that the spirit of the sabbath is more important than the letter of the law. When Jesus' disciples husked some corn and ate it on the sabbath, they were accused of breaking the rule of rest. But Jesus answered 'The sabbath was made for man, not man for the sabbath' (Mark 2:27). That is to say that the rules about the sabbath should not be regarded as laws laid down from outside. Rather the sabbath should be part of a whole way of life which includes honouring God and respecting his creation.

The responsibility of women and men, then, was to care for the household and the land in a way that would cultivate wisely the creation that God in his wisdom had made. That was part of their relationship with God—the relationship which gave them their identity. Time and time again, the stories of the Old Testament stress that if people went their own way, ignoring God, evil results would occur, not only among humans but also between humans and the rest of creation. The very earth would suffer from human folly and lose its productiveness as part of God's judgement. This repeats the message of the second creation story: obedience to God maintains the fruitfulness of creation; disobedience brings human exile and turns fruitfulness into wasteland. All creation's well-being is interrelated, but it is humanity's responsibility to maintain that well-being.

When the prophets came to speak to the people of Israel about their failure to live in relationship with God, they often used the analogy of plants. The people knew about the link between themselves and the natural world, and they understood: 'Your mother was like a vine in the vineyard transplanted by the water, fruitful and full of branches', wrote the prophet Ezekiel, but now 'the vine was plucked up in fury, cast down to the ground. . . . Now it is transplanted in the wilderness, in a dry and thirsty land' (Ezekiel 19:10–13). Ezekiel was speaking to the people of Israel at a time when they had been conquered and taken into exile, and he was telling them that this was because of what they had done wrong.

According to many prophets Israel had gone wrong in two ways. It took too little notice of God, and it treated the poor unjustly. 'They sell the righteous for silver and the needy for a pair of shoes', accuses Amos (Amos 2:6). Amos also shows how God's anger towards Israel has effects on the land: 'the pastures of the shepherds mourn, and the top of Carmel withers' (Amos 1:2). Yet when Israel has been thoroughly sifted and shaken out of its injustice, the land too will be restored: 'they shall plant vineyards and drink their wine and they shall make gardens and eat their fruit. I will plant them upon their land and they shall never again be plucked up' (Amos 9:14). The same sequence occurs in Isaiah. His famous vision of restored harmony when 'the wolf shall dwell with the lamb and the leopard shall lie down with the kid' will happen only when a man in the spirit of the Lord will judge the poor with righteousness and decide with justice in favour of those who are humble. When righteousness and faithfulness produce this justice in the land 'the sucking child shall play over the hole of the asp, and the weaned child shall put his hand on the adder's den.

They shall not hurt or destroy in all my holy mountain; for the earth shall be full of the knowledge of God as the waters cover the sea' (Isaiah 11:8–9).

Questions for discussion

1 Why is the well-being of the land seen as being so closely bound up with the behaviour of human beings?

2 Can you point to examples today where human injustice, wrongdoing, or going against God have resulted in ecological damage?

Yet the Old Testament also raises questions about this link between obedience to God, prosperity, and the well-being of the land. The most serious questioning comes in the book of Job. Job is a good man who loses, one by one, his farm animals, his many sons and daughters, and his health. Three of his friends gather round him and try to persuade him that he must have done something wrong to deserve such a terrible fate, but he insists that he has always served and obeyed God. Why then does he suffer? The question is never answered, and the book closes with a thundering and powerful declaration of God's power and the wonderful creatures he has made, while Job has to admit that he knows nothing, and can only bow to God's overwhelming power and wisdom.

The book of Job is part of the group of books in the Bible known collectively as the Wisdom tradition. They include the books of Proverbs, Psalms and Ecclesiastes, all of which return to the same questioning (why do good people suffer?) and the same declaration of God's power and wisdom. Wisdom herself is spoken of as 'she', and personifies that spirit that comes from God and directs women and men into right living. She is pictured as having been created before anything else and 'like a master workman' was with God 'when he marked out the foundations of the earth' (Proverbs 8:23, 30). By setting up her house among men and calling them to come and learn from her, Wisdom personifies the character of God in orderly creation. The writers of the Wisdom books may not be able to account for why the wicked flourish in God's orderly world, but they continue to have reverence for the God who made it all.

To sum up thus far: creation in the Old Testament is the orderly, beneficent work of God, a cause of wonder, delight and knowledge to the poets who wrote the Psalms, a cause of hard work to farmers, yet still able to be productive as God from the first intended so long as humanity lives and believes as it should. All dealing with the natural world takes place within the covenant with God and therefore has to be conducted with respect and an ecological soundness which decrees rest and the conditions necessary for fruitfulness.

Question for discussion

'Creation . . . is still able to be productive . . . so long as humanity lives and believes as it should.' Do you think that what people believe as well as how they live affects the well-being of the created world?

THE NEW TESTAMENT

In contrast to the Old Testament, which Christians share with the Jews, the New Testament is specifically Christian, and tells of the birth, life and death of Jesus, the life of the early Church, and the deeper meaning of Jesus' life and death. But it does not represent a break with the traditions of the Old Testament, rather a new building on that tradition.

Jesus was a Jew, and grew up in a rural community which was part of that tradition, so it is not surprising that his sayings and parables reflect it. His stories describing the Kingdom of God are often set against a farming or fishing background, vividly illustrating God's presence in everyday life. His stories tell of weeds being sown amongst wheat, of seed which fell on good or poor ground and so gave a good or poor harvest, or of fishermen sorting their catch (Matthew 13:24–30, Mark 1:1–9, Matthew 13:47–50). All these stories show the contrast between waste and fruitfulness. Jesus' famous story of the shepherd who goes to great lengths to search for one sheep that is lost takes respect for animals

31

as the basis for understanding God's care for humanity. At other times, still teaching about God's care for those he has made, Jesus could point to the lilies of the field (Luke 12:27) or the birds of the air (Matthew 6:26). If these brief and fragile lives flourished in God's care, why should human beings be so worried about saving money and possessions for the sake of security? Indeed one effect of following Jesus' teaching about the Kingdom of God is that one lives lightly upon the earth, not creating surpluses for one's own comfort (Luke 12:16–21) or weighing oneself down with riches (Luke 8:14).

Question for discussion

From the examples given above, and from your own reading of the New Testament, what do you think was Jesus' relationship with and attitude to the natural world?

In the Old Testament the main theme is the relationship between God and the people of Israel. In the New Testament the main theme is the new thing God was doing for humanity, through Jesus Christ. The Old Testament theme of relationship included the rest of creation within its scope, especially as Israel was largely a rural, agricultural people. But the New Testament, apart from Jesus' time in Galilee, is concentrated much more on urban groups hearing and living the Gospel, so there is correspondingly little about the natural world.

The New Testament contains descriptions of Jesus in his life, an account of how his first followers began to teach that he was alive and active in the world through his spirit, and letters written by these early Christians which explain the faith to new converts, and encourage them to live by that faith. Foremost among these first Christian teachers was Paul, who never knew Jesus in his life, but who had an overpowering vision of Christ while on the road to Damascus, which changed him from a hater of Christians to a believer in Christ. Much Christian thinking about the meaning of Jesus' death and resurrection, and about Jesus' place in God's Kingdom, comes from the explanations given by Paul in his letters.

For example, Paul or someone influenced by him wrote to the new Christians in Colossae that Christ 'is the image of the invisible God, the first-born of all creation; for in him all things were created, in heaven and on earth . . . all things were created through him and for him' (Colossians 1:15–17). This hymn of praise echoes some of the Wisdom writing in the Old Testament which we looked at on pp. 30–1. There, Wisdom is described as working with God in creation, and showing people on earth how to live orderly lives in creation. Here in the New Testament that idea is applied to Christ: everything, all creatures, are 'in him . . . through him and for him'. So Christ is seen in two ways which come together to make the whole picture. He is the perfect man who died to bring the world back to God, and he is divine—'the image of the unseen God'—whose word created everything that exists. Everything, from first to last, exists 'in Christ'. In the Old Testament, human beings are told to manage the created world as representatives of God's dominion. In the New Testament human beings are also called upon to love the created world, as representatives of Christ's love.

This means that Christ expresses the whole of God's love in every place for all the inhabitants of the world. Therefore no creature in any circumstance is merely, say, an animal, an insect or a bird. Every creature is 'in him . . . through him and for him' and so is worthy of our wonder, respect and loving concern. With this understanding of the natural world having divine as well as practical importance, practices like the battery farming of hens become intolerable.

Human beings in their earthly actions are to be transformed into the image of Christ, which can also serve to express final hope for the whole creation. In the fullness of time God plans to 'unite all things in [Christ], things in heaven and things on earth' (Ephesians 1:10).

The Jews looked forward to a transformation that would take place at the end of time, when everything would be changed into a new heaven and a new earth, when all the sin and grief of this world would be forgotten. This hope was not just for human beings: 'The wolf and the young lamb will feed together, the lion eat straw like the ox, and dust will be the serpent's food. They will do no hurt, no harm on all my holy mountain, says the Lord' (Isaiah 11:6, 9). Paul came from a Jewish background, and inherited this tradition. At the moment, he says, creation is 'in bondage', groaning as if in childbirth, because humanity has been disobedient to God, and the rest of creation shared our punishment. But there is hope. Paul believed that this transformation, for human beings and all other creation alike, had been made possible by Christ's resurrection, and that everything was rapidly moving towards it.

The last book of the Bible, the Revelation of St John, also describes a vision of a new heaven and a new earth (Revelation 21:1). There at last would the order, harmony, fruitfulness, justice and knowledge of God which had been intended by God for creation have its full realization.

Questions for discussion

1 If every creature is 'in Christ, through him and for him', what response does this evoke in you? Are there some parts of the natural world that you find difficult to include in this?

2 When human beings are 'transformed into the likeness of Christ', what will be their relationship with the natural world? Give some examples of how they will live.

3 THE INFLUENCE OF THE BIBLE ON CHRISTIAN BELIEF ABOUT THE NATURAL WORLD

Dr Ruth Page

For Christians the Bible is the record of Israel's experience of God and the early Christians' experience of God in Christ, and Christian belief is based firmly on the Bible. At the same time, as we saw on p. 23, the Bible is a series of books written by many authors over hundreds of years, and each book reflects the situation, viewpoint, beliefs and problems of the writers. So the Bible is the way God speaks to his people at many different times and places, but it is also a number of very human books written at particular places and times. Christians have to accept that the Bible has this double nature. If God spoke to us in his own terms, we would not be able to understand. Knowledge of God must therefore be expressed in comprehensible *human* terms.

But this is a changing world, so the human terms which were clear and appropriate at one time may become unclear or inappropriate at another. Biblical writers were in the first place speaking to their own day and generation, so we have to try to understand what they were saying at that time, rather than taking out isolated verses and forcing them to fit a new pattern of 'the Christian viewpoint for today'. If we do try to understand the Bible on its terms rather than our own we can actually see more clearly where it does indeed give us guidance for today, where it has been misunderstood by Christians, and where the conditions of life have changed so much that what the Bible says can hardly help us with present-day problems.

One clear example of that last category is the command given to men and women in the book of Genesis to 'be fruitful and multiply', a command often repeated in Christian writing. The context of these words in Genesis 1 is a picture of a newly-created, empty world which the first creatures are to fill. But the same values are found all through the Old Testament. Barrenness, lack of children, and wasteland were a cause for grief and were often held to be a sign of God's displeasure. It is understandable that in the lightly-populated world of the original writers, where wilderness was a threat rather than a delightful difference from urban sprawl, attitudes would be different from today's. The joy of a new baby, of beautiful gardens and productive farms is still with us. But our world is so much more crowded.

Paul and Anne Ehrlich wrote in *The Population Explosion* (Hutchinson, London, 1990, p. 9): 'Each hour there are 11,000 more mouths to feed; each year more than 95 million. Yet the world has hundreds of billions *fewer* tons of topsoil and hundreds of trillions *fewer* gallons of groundwater with which to grow good crops than it had in 1968.' They plot in terrifying detail the precarious food situation of the developing world, the problems of agriculture as population grows, the dangers of pressure on the health of global ecosystems and other issues. In the light of all that we cannot simply follow the command to be fruitful and multiply. Even when the birthrate in developed countries falls, one baby in such lands of plenty puts more strain on the life-support systems of the planet than many babies in a poor country. The spirit of the original command was that creation was bountiful. If we are still to experience the bounty of creation we must not exhaust its resources by blindly following this command in Genesis.

Another related area in which our circumstances are very different from those in which the Bible was written concerns the unqualified approval given to the cultivation of land. The Jews gave thanks that the dry, stony soil of Palestine had some fertile areas, and could be tended and irrigated to make it fruitful. This was a wonder of creation. In areas bordering on desertification we may still give thanks for any successful farming. But in the process of time biblical praise for farms and vineyards gave rise to the view that cultivation was the only really worthwhile relationship with the land or the seas. This opened the way to over-cultivation. The attitude involved is seen clearly in Knut Hamsun's novel *Growth of the Soil* (Picador, London, 1980) where one man goes into the Norwegian forest and clears by hand an area for his first field and farmhouse. Gradually through his struggle and back-breaking work, his success and disaster, the farm grows and the trees around are all felled. More people move into the area and a whole settlement begins with a road to it through the forest. Towards the end of the novel the first tractor appears. When this first came out (in English in 1921) it was thought a great account of human relationship with nature. It becomes more understandable in the Norwegian context of the looming presence of great forests, but today it reads as one more account of human self-assertion over and against nature. It is told as the worthy labour of a man bringing land into cultivation, but there is no thought for what might have been disturbed or lost in the process. It is only recently that we have come to see that enough is enough in the way of 'development', that forests are precious, that wilderness and wetlands are to be preserved not just as something for poets to gaze at but as habitats for species which are themselves valuable. The Old Testament valuation of cultivation over wilderness was right for its time and limited activity, but a different view is required now of land and sea, the resources we all share and which once seemed plentiful but which are now rapidly decreasing.

Questions for discussion

1 How should the biblical command to 'be fruitful and multiply' be practised now?

2 What do you understand by the word 'fruitful'?

3 There are very few truly untouched wildernesses left in the world today. What is the wildest place you have ever seen? What was your reaction to it? What do you think the Old Testament writers would have thought of it? What do you think is God's view of it?

The Old Testament is full of practical and wise concern for the useful plants and animals of farms and vineyards. That is partly utilitarian, i.e. making sure that what is useful remains useful, but it is not simply materialistic because everything was included in the covenant with God. In the Psalms, though, the Old Testament does go beyond what is useful and celebrates wild nature as well as tame. But the Psalms are poems, and it is not clear whether the attitudes expressed in them were ever put into practice.

There has been a long theological tradition in Christianity of creation's 'plenitude', which originally meant that God had made just the right amount of variety in the species on earth. Since it is now known that species evolved rather than were 'made', and that many have died out, plenitude can no longer refer to the completeness of God's creating action. But it can still express the richness of diversity of creatures wild and tame, beautiful and ugly, useful and useless which remains a wonder. What is new in our thinking is that efforts for conservation are no longer only for useful species but in order to preserve that diversity and variety.

Useful species are useful to *humanity* and there is always a danger that what is not useful or appealing will be given no consideration. This is discussed at length in Freda Rajotte's article, pp. 7–8. This putting humanity at the centre can be seen in the Bible. Care for other *people* is commended—even people who might not be noticed, like the poor, who aroused Jesus' compassion. Compassion for creatures, however, is rarer, and although there have always been some Christians who practised it, the omission has developed over the centuries into an unconscious assumption that humanity's desires and comfort are more important than anything else on the planet, as we have seen earlier. In Israel, God's covenant with the people was far more central than God's creation of the world. When Jesus spoke of how God cares for lilies and ravens he went on to say how much more God valued humanity. There are, though, some examples of the opposite view in the Bible. In the book of Job, Job is overwhelmed by a presentation of the sheer variety of God's creation in relation to which—not only in relation to God—he is 'of small account' (40:4). The selfishness in seeing everything in relation to human beings is balanced by the realization that all God's creatures matter to God so that, for instance, every decision on land-use involves consideration of the species at home there who cannot present their own case. We may also remember that just as in the Old Testament *nephesh*, breath, was something shared by humans and animals, so today we know that our genetic material is not very different from that of other species. These are our fellow creatures.

Questions for discussion

1 Can you reconcile the idea that God made the world perfect, with the evidence that species have constantly evolved and died out in the course of evolution?

2 Many appeals on behalf of the environment and endangered species concentrate on the harm we are doing ourselves by harming the environment. Do you think this is:
   (a) natural and right?
   (b) maybe a bit limited, but a good way of engaging people's interest and support?
   (c) a wrong and harmful view?

Christianity and Judaism have been accused of worse than anthropocentricity, however. From the verse in Genesis (1:26) in which God granted 'dominion' to the first man and woman it has been argued that these religions turned dominion into domination and are responsible for the degradation of land and for the exploitation of resources to the point of their exhaustion. This point is dealt with from different angles in other articles in this book. I have already argued (p. 25) that this verse in Genesis does not suggest exploitation, for it describes humanity acting as God's representatives in a world which has just been created harmonious and fruitful. The accusation itself may be overstated, for exploitation of the environment began in a small way with the first human groups and has been most thorough and uncaring when Eastern Europe was Communist. What may have come through Judaism and Christianity is the expectation that land will be useful and is to be used. Thus the Puritans who settled in America in the seventeenth century believed that they were using land as God meant it to be used when they began farming more thoroughly than the native Indians who hunted and gathered as well as doing some farming. More recently, in a society where God began to seem less important, the notion remained that the land and the waterways

39

were there for human use and could be exploited selfishly, and again no voices from Christianity were raised in protest. In one way the ecological crisis has made humanity extend its idea of usefulness: trees, for instance, are useful for trapping carbon dioxide, a greenhouse gas; all manner of surprising plants may be useful in medicine. But if dominion is never to be domination, we must realize that the creatures of this world in all their interrelationships are much more than the instruments of human beings.

It has been suggested that one reason that forests have been cut down so freely, mines and wells exhausted, and dangerous chemicals released into the atmosphere is because the Bible does not describe God as living within creation, so that nature is not seen as sacred. There is some truth in this, although again much of the worst destruction was carried out by men who did not hold any religious belief, for whom profit was a more vital concern than religion.

To see where there is some truth in this accusation we have to return to the ancient world of many gods close at hand, often competing with each other, who had all to be kept happy. Israel became a nation at this time, and therefore had to be clear about the differences between its God and those around it. The Israelites described one eternal transcendent God at work in creation. Now we are not surrounded by large numbers of gods all demanding attention and threatening disaster, so we can see more clearly the problems which can arise from certain features of the biblical account of creation. In Genesis 1 God is so transcendent as to be remote from creation. The older account in Genesis 2, in which God works with clay, is concerned for Adam's solitude and walks in the garden, describes better a God with whom one could have a relationship. But in both cases it is clear that God does not live in creation like the gods of other nations. Christianity continued to emphasize God's transcendence, since the Creator was not to be confused with, or enclosed by, the creation. Therefore wherever Christianity spread, local gods and goddesses inhabiting forests, streams and wells were officially denied, even if unofficial belief in them lingered on. Thus the natural world was 'de-divinized', i.e. all sacred presences were removed from it and it was left virtually secular. Israel learned its faith through history, and it was in history rather than in nature that Israel, and also Christianity, expected to find God at work, although that work had effects on nature.

Part of the problem is that transcendence has been thought of in spatial terms, as a very great distance 'up'. Psalm 139 does speak of God's presence everywhere, but more often God is thought of as being above the sky. In that case a transcendent God is certainly remote and although God's Son and Spirit might be evident among *people*, it was not often thought that God had concern for the lives and disasters of creation other than human beings. Moreover, later Western theology was mainly based on urban society and although the right use of nature was encouraged—by, for instance, St Benedict in his monastic rule—there were very few who saw, as St Francis did, that *all* creatures are our brothers and sisters under God our Father. The Old Testament is admirable in recommending respect for species that were used and even eaten. Today we can see that the same respect is due to all creatures, even when difficult decisions involving making priorities among them have to be made. Notions of God's transcendence can give way to an understanding of God as being present everywhere. The God who is everywhere is present with and caring for all creatures great and small—a matter which will greatly increase our respect. This is not a return to a number of little local gods. It is the sense which has flashed on Christianity from time to time, of the world being permeated by God's loving, stirring, confronting and consoling presence with, but not absorbed into, creation.

Questions for discussion

1 How do you picture God to yourself—above your head, beside you, inside you? How do you think this affects the way you think about God's relationship with and concern for the world?

2 In what sense can the world be said to be filled with God's presence?

Is this a 'good' world? The Bible, especially in the Psalms, says that it is but can also imply that it is not, as when Paul describes creation 'in bondage' and 'groaning' in Romans 8. These two different answers have led to two different emphases in opposing streams of theology, one saying that the world is basically good in spite of humanity's disobedience and punishment, and the other seeing the whole creation fallen into disobedience of God's will. Both views in fact are one-sided, for the world has probably been a mixture of better and worse for most of its inhabitants. It is too simple to appeal to the repeated phrase of Genesis 1 that 'God saw that it was good' as a religious reason for opposing ecological vandalism. Rather than asserting that the world is good and must therefore not be destroyed, it is better to point to God's concern for creation.

Commentators on the Bible seem to agree that in its context 'God saw that it was good' meant that the creatures which were brought into being were fit (suitable) in themselves and fitted into an overall order. This order, moreover, was vegetarian—meat-eating was introduced after the Flood—and since the writers believed that the plants lacked *nephesh*, the breath of life, no violence was being done. This is not saying that the world is wholesome or morally blameless if left alone by humanity, but that God created order in which everything had its place, not interfering with anything else. The Priestly writers of the sixth century BCE were probably describing a totalitarian orderliness given by God. That was what *they* appreciated. That was 'good'.

We now know what the biblical writers could not know, that the diversity of species came about through the long, untidy history of evolution. Religious responses to ecological disaster which do not take evolution into account are not meeting present issues because it is evolved ecological systems which are at risk— in fact, to echo Genesis 1, it is the continuation of evolved *order* which is endangered. Goodness of the creating process in any other sense, however, is hard to affirm, especially in the light of the huge number of species which failed to survive long before humanity began to make its impact. Genes may have continued from species to species, but individuals and species themselves were part of a food chain (unlike Genesis 1) and were always vulnerable to change in their circumstances. Thus the dinosaurs, the most dominant species for millennia—far longer than humanity has been—were wiped out, probably because of a climatic change to coolness which the solar heating of their bodies could not adjust to.

God, surely, cannot be said to have wiped out species after species. Yet if God is as much in control of the world as the Old Testament frequently suggests, that would have to be the conclusion. To deal with this difficulty a theological point often made about humanity has to be extended to creation as a whole. It is often said that God gave humanity freewill so that people would not be puppets. When men and women turn to God, in that case, they do so freely. But the gift of that freedom carries with it the freedom to fail, whether in responding to God or in loving one's neighbour. The human moral evil of the world is thus explained by the misuse of human freedom. This can be extended to the natural evil which occurs in such things as earthquakes or the natural extinction of species. Thus instead of God having made all creatures in order at one go, one may think of God 'letting be', giving creation freedom to develop as it could, but having concern and love for all that did develop. The development was not a triumphal progress but a varied process in which some genes and species survived and others died out. But the value to God should not be thought of in the long view of the whole process (we do not think that way of history) but in each life that was lived, which was valuable to God in its time. But in such a dynamic, changing world we cannot easily say that it is good—nor, for that matter, evil.

Thinking about evolution, moreover, pricks the bubble of our human self-importance. In the Christian tradition people have tended to see human beings as very important in creation because they are made in the image of God. Evolution helps us put that in context. In his *Life on Earth* (Collins/BBC, London, 1979, p. 20) David Attenborough plots the whole course of evolution on the timescale of a year. On that reckoning the first humans did not appear until the evening of 31 December. For almost all of its life, then, creation has lived without humanity. But God would not abandon the world; rather God would have the right kind of relationship with all the creatures who evolved. Humans tend to picture God in terms of another kind of human and therefore imagine him having a human and limited relationship with plants, rocks or animals. But that is to limit God who understands both humans and the rest of creation deeply and fully. Humanity, therefore, that newcomer in the history of the world, cannot be the divinely-appointed mediator between God and the rest of creation as if creation and God could not be directly related. One may also be horrified at how much damage this new human arrival has been able to do in so short a time. That does not deny humanity's intelligence and ability, nor its responsibility before God for using its freedom and ability for the well-being of creation. But its capacity to image God is to be sought precisely in fulfilling that responsibility.

In the Bible obedience to God in exercising responsibility wisely is often said to be rewarded with productive land, children and long life. 'The righteous flourish like the palm tree, and grow like a cedar in Lebanon' (Psalm 92:12). Disobedience brings exile, either from the garden of Eden or from the Promised Land, while the land itself suffers harm. This pattern is again part of the neat orderliness of the theological view of creation. It is questioned even within the Old Testament itself: 'Why do the wicked live, reach old age, and grow mighty in power?' (Job 21:7). Things are certainly not as tidy as the simplest teaching of the Old Testament would have it. Yet there remains a point which is clearly seen in these days of global warming and a hole in the ozone layer. Wickedness towards the earth does have its consequences. The earth is not an endlessly self-renewing system for human consumption and waste. Individual wealthy exploiters may have died peacefully in their beds at a great age, but we are now collectively suffering the accumulated effects of their wickedness (the biblical word). Righteousness in this case is leaving the world at least no more polluted than when one entered it. There may be no equation between that and individual prosperity, but there is an equation between collective righteousness and the good of the world.

Questions for discussion

1 Is the created world good?

2 What exactly does 'conservation' set out to conserve?

3 What is humanity's responsibility before God with regard to the natural world?

Thus far I have been considering the theological implications of what the Bible has to say on the natural world in the light of the ecological crisis, sometimes critically in view of our current problems, sometimes appreciatively. But lying behind all the particular points discussed are two beliefs which are fundamental and continue to be affirmed.

The first is that this world, and the universe in which it is set, is the creation of God, by which it is given meaning and purpose. Descriptions of God's actions and the actual way the world began may vary, but that is only to be expected. There are three in the Old Testament—creation by word, by modelling clay and God's Wisdom using geometrical precision. The early one was kept in the Jewish tradition as the later new ways of understanding emerged. Such development of belief without throwing out the old is the sign of a living faith which continues to respond to circumstances. As Christianity is still a living faith responding to circumstances the ecological crisis will have an impact on belief. Astrophysics, chaos theory or the theory of evolution may contribute something to our understanding for our own day also. But the character and the will of God who let it all be remains. To Christians, the world cannot be explained only in its own terms.

The second fundamental inheritance from the Old and New Testament is that life is to be lived in accordance with this belief. In the Old Testament that was expressed in terms of a covenant between God and Israel which covered all aspects of life including justice for the poor, as the prophets insisted, and consideration for creatures in the charge of humanity. In the New Testament Paul wrote of being 'in Christ', an expansive phrase which includes finding one's environment changed by belief in Jesus Christ and being in the fellowship of 'the body of Christ'—other Christians. Although there are no specific recommendations for Christian living in relation to the natural world, to follow Jesus in that regard is to make no heavy demand on natural resources, to enjoy the world and find in it parables of relationship with God—just as

Jesus spoke of the shepherd's care for the sheep, or the way the seed grew in the ground. Paul indeed extends the notion of salvation to the whole of creation which in the end is to be released from its bondage and will 'obtain the glorious liberty of the children of God' (Romans 8:21). Thus all of creation, the whole natural and human world shares in the hope expressed in both Testaments out of a situation of difficulty, pain and near despair. There will be in the end a new heaven and a new earth where the harmony of creation in the presence of God will be complete. Harmony arrived at now in the midst of all the problems of earth is an anticipation of that future state.

> Jesus said therefore, 'What is the kingdom of God like? And to what shall I compare it? It is like a grain of mustard seed which a man took and sowed in his garden; and it grew and became a tree, and the birds of the air made nests in its branches.'

> (Luke 13:18–19)

Questions for discussion

1 'To follow Jesus is to make no heavy demand on natural resources and to enjoy the world.' What are your enjoyments? How much demand do they make on the natural world? Could you change the way you enjoy the world to make it more consistent with Jesus' example?

2 'Harmony arrived at now in the midst of all the problems of earth is an anticipation of that future state.' What progress towards harmony can you see in the world today? Give some examples, and say how they might be pointers to the future 'new heaven and new earth'.

4 | PRESERVING GOD'S CREATION

Greek Orthodox Metropolitan John of Pergamon

(*Adapted by Elizabeth Breuilly from lectures given at King's College, London, January 1989, and published in* King's Theological Review)

It is becoming increasingly clear that what has been named 'the ecological crisis' is perhaps the number one problem facing the world-wide community of our times. It is a global problem, concerning all human beings regardless of where they live or their social class. It is a problem that is not simply to do with the well-being of humanity but with the very being of humanity and perhaps of creation as a whole. It is difficult to find any aspect of what we call 'evil' or 'sin' that is so all-embracing and has such devastating power as the ecological evil.

In view of this situation what does Christianity and its theology have to offer to humanity? First of all we must say that theology cannot and should not remain silent on an issue like this. If faith is about ultimate things, about life and death issues, the environmental problem certainly belongs in that category. Christian theology and the Church should not have stayed silent for such a long time on this matter. Particularly since they have both been accused of having something to do with the roots of the ecological problem—and there is some justice in this accusation (see Chapter 1). The Church and theology have to speak on this matter not so much in order to apologize or defend themselves against these accusations, but in order to offer their constructive contribution to the solution of the problem.

When our Western societies think about possible solutions to the ecological problem, all our hopes seem to be placed in *ethics*—the study of right and wrong. Whether enforced by state legislation or taught and instructed by Churches, academic institutions etc., the hopes of humankind seem to be based on ethics. If only we could behave better! If only we could use less energy! If only we could agree to lower a bit our standard of living! If, if . . .

But people do not give up their standards of living because this would be 'rational' or 'moral'. By appealing to human reason we do not necessarily make people better, and moral rules seem to be more and more meaningless and unpleasant to people nowadays, especially since these rules are now largely separated from religious beliefs.

Human beings do not always behave rationally and cannot be made to do so either by force or by persuasion. There are other forces, besides the human intellect, that decide the direction in which the fate of the world moves. Theology and the Church ought to embrace areas other than rules about right and wrong if they are to be of any use in this case. These areas must include all that in earlier times used to belong to the mythological, the imaginative, the Sacred.

If we try to solve the ecological problem by introducing new ethical values or rearranging the importance of the traditional ones, I fear that we shall not go far in reaching a solution. So what is the answer? Later on I shall put forward some ideas which aim at building an ethos—an atmosphere, a way of seeing things, which is based on some of the earliest traditions of the Church as well as on present-day perceptions of the world and humanity's place in it.

Questions for discussion

1 'People do not give up their standards of living because this would be "rational" or "moral".' Do you change the way you behave because of convincing arguments? If not, what might cause you to act differently?

2 Do you see the ecological crisis as an aspect of evil or sin?

A GLANCE AT HISTORY—1. THE FIRST CENTURIES

First let us look at how ideas about the natural world have developed since the birth of Christianity.

Christianity took shape in the context of two cultures, the one dominated by the Jewish and the other by the Greek way of thinking. In what way did these two cultures understand humanity's relationship to nature, and the place that God occupied in this relationship?

The Jewish way of thinking tended to give importance to *history* (the history of the Jews, the chosen people of God in particular) and to see God as revealing himself mainly in and through his acts in history. Nature was a less important way in which God revealed himself, and at times even this was looked at with some suspicion, because any sort of reverence for nature, or any importance attached to it carried with it the danger of falling into nature worship. This could not be allowed, since what made the people of Israel special, what gave them their identity, was their allegiance to the one God who created nature, but did not dwell within it.

Greek culture, on the other hand, attached little importance to history. Nature offered to the Greeks the sense of security they needed, through the regular movement of the stars, the cyclical repetition of the seasons, and the beauty and harmony which the balanced and moderate climate of Greece (at that time) offered. The study of the universe was the main concern of the Greek philosophers. They saw God present and operating in and through its laws of cyclical movement and natural reproduction.

These two views, which lie at the heart of Christianity, give rise to two very different understandings of the created world. For the Greeks the world was a reality which contained its own energy, its own power to survive. It would exist for ever because that was the way its systems worked. By its very nature it was eternal.

The Jewish view, on the other hand, saw the world, the whole created universe, as an *event*, a gift from God. Its continued existence from moment to moment depended on God, and it could be relied upon only because God could be relied upon.

The early Church lived in both these cultures, and it had to combine a world-view that trusted nature for what it was and believed that it operated according to its own laws, and one that regarded it as a gift and an event, constantly dependent on its Creator and Giver.

The marrying of the two views gave rise to a view of the world as good and beautiful and as an important concern of human beings. But in this view all this depends not on itself but on God. In order to enjoy the world, to study it and rely on it, humanity must refer it back to something that is outside and beyond the world—that is, to God. Thus, the earliest Eucharistic prayers of the Church involved a blessing over the fruits of the earth, but also an affirmation of faith in the survival of creation and nature. So creation and nature, and not just human beings, were central in the Church's thinking.

To sum up this point, the Christian view which emerged from the encounter between Hebrew and Greek thought, is that the world is an *event* and not a self-explainable process, but that it can be said to be permanent and to survive because of another *event*, namely its being referred to the eternal and unperishable Creator.

So in early Christianity interest in nature and in the cosmos held a central place in the Church's thinking, but this did not fall into paganism because it was seen in terms of two continuous events: God's constant giving or sustaining of the world, and the world's being constantly referred back to God. But who was to refer the world to God? The inanimate world could not do this for itself. This is where the special role of humanity comes in, and we shall look at this again later on.

Questions for discussion

1 Think about your own life and your own experience. Think of one example of how you have learnt more about God through events, through something that happened. Think of one example of how you learnt more about God through the natural world.

2 In what ways can the natural world be said to be constantly dependent on God?

3 Do you feel that you can rely on the natural world?

A GLANCE AT HISTORY—2

In the West a tendency developed to introduce a distinction between humanity and nature by regarding humanity as superior to nature, and as the centre of everything. Early writers such as St Augustine and Boethius defined the human being in terms of reason, intelligence and the ability to be aware of oneself. They described God in the same terms, as having a greater reason, a more profound intelligence than human beings. Human beings were singled out from all God's creation as being not only a higher kind of being, but in fact the only being that mattered in eternal terms. According to St Augustine's view there is no place for nature in the Kingdom of God. That Kingdom is only to do with spiritual beings, with eternal souls. The Church was gradually losing its awareness of the importance and value of the physical created world.

All this depends on how you see the special nature and role of humanity. The Bible tells us that God created human beings 'in his own image'. But what does this mean? There are clearly so many ways in which we are *not* like God, but we must be like him in some way that is not shared with the other animals. The term developed to discuss this question is the Latin phrase *imago Dei* (image of God). The *imago Dei* is whatever quality it is that makes human beings, and only human beings, an 'image of God'. There has been much discussion over the ages about what exactly the *imago Dei* consists of.

The Middle Ages and the Reformation gradually reinforced the idea that the *imago Dei* consists in human *reason*. In Western Christianity the sacrament of the Eucharist or Holy Communion no longer centred on offering back to God the good gifts of his creation, but concentrated on the souls of the believers who took part in it. The sacrament became irrelevant to the material world, and the gap between nature and humanity widened even further.

In the centuries that followed, Western religion, philosophy and literature all combined to reinforce the idea that human consciousness, human thought and human action were the centre of the created world. Even movements such as Romanticism, which

51

celebrated nature, underlined the distinction between the thinking, conscious observer and non-thinking, non-conscious nature. Nature only had value when a human being observed it, reflected on it, and reacted to it. Religious and theological movements still operated without any reference to nature, while Puritans and mainstream Calvinism exploited to the utmost the Genesis verse urging us to 'multiply and dominate the earth' (see Chapter 7), thus giving rise to capitalism and eventually to technology and to our present-day civilization.

The first challenge to this human-centred and reason-dominated world-view came not from within Christianity, but from Darwinism. Darwin pointed out that the human being is by no means the only intelligent being in creation and that consciousness, even self-consciousness, is to be found in animals too, the difference between them and human being one of degree not of kind. This was a blow to the idea that the *imago Dei* consists of reason or intelligence.

Thus, humanity was put back in its place as being part of nature, and the question had to be posed again: what constitutes humanity's difference from the animals, since we now see that reason is no longer the *special* difference? The Church failed to respond constructively to the challenge of Darwinism. It either continued to insist that reason was the *imago Dei*, and entered into battle with Darwinism, or it accepted the view of humanity as only another animal, and refused to look for other areas that make human beings different.

But Darwinism has virtually won the science of biology for itself, and will not go away. Theology has to make the best use of it—both positively and negatively—if there is to be any hope of overcoming the ecological crisis.

Not only biology, but other developments such as quantum physics, are making us review our traditional theology. I believe that this pressure can be of decisive benefit to the Church in its attempt to face the ecological problem. In order to do this, however, the Church must make a creative use of all the new developments in the areas of biological and natural sciences, together with an exploration of the Christian tradition to find elements which can be built into a new understanding. Our modern world has passed through changes that make a return to the past both impossible and undesirable. Theology today must use the past with respect, and learn from it, as I have tried to show above. But theology must try to adjust the past to the present by creatively combining it with whatever is best in science, art, philosophy, and other areas of thought.

Questions for discussion

1 What do you see as the difference between human beings and all other creatures?

2 How do you understand the idea that God made human beings 'in his own image'?

3 Do you think that God is more concerned with the souls than with the bodies of human beings?

THEOLOGY FOR THE SURVIVAL OF THE WORLD

We have seen how the Christian Church in the early centuries viewed the world as God's creation. On the one hand it was stressed that God the Father created the universe freely and out of love. On the other hand it was stressed that God created the world out of nothing. He might have chosen not to do so, so the existence of the universe is a free gift, not a necessity. From this view it follows logically that creation is under constant threat of return to nothingness, a threat which individual beings experience as decay and death.

But Christian faith goes hand in hand with hope and love. Knowing what God has done for the world, we know that he loves it—that he created the world out of love. So there must be hope for the world's survival. But how? A simple answer might be that since God is almighty he can simply order things to happen so that the world may survive in spite of what is done to it. In other words, God could work a miracle to save the world. Perhaps this is the answer given by most people in the face of threatened disaster. But Christian faith does not believe in this sort of mechanical solution. In some ancient Greek and Roman tragedies, when everything seemed to be moving to disaster with an inescapable logic, the solution was to lower a god onto the stage from above, who would then set everything right and undo the damage done. This came to be known as the *Deus ex machina*— the god from the machine. History, the Bible, and our own experience teach us that our God does not act in this way.

But in creating the world God did not leave it without the means for its survival. In creating it he provided also for its survival. What does this mean?

In spite of all human shortcomings, and the damage that humans have done to the world, I want to show that the solution of the problem lies in the creation of humanity. How and why do we believe humanity is capable of performing such a role?

Questions for discussion

1 Christianity teaches that God created the world and that God loves the world. What are your reasons for believing this, or not believing it?
2 Do you feel that 'creation is under constant threat of a return to nothingness'? Does this worry you?

WHAT IS HUMANITY?

We looked earlier at the idea that the human being is superior to the rest of creation by virtue of having reason. Humanity's task is seen as being 'to interpret the books of nature, to understand the universe in its wonderful structure and harmonies and to bring it all into orderly articulation'.

This view, that humanity's distinctive identity and role in creation is because of our rationality, has contributed a great deal to the growth of the ecological problem. For reason can be used in both directions: it can be used as a means of relating creation to the Creator in an attitude of praise, but it can also be used as an argument for turning creation towards human beings and human ends. And that is the source of the ecological problem.

Humanity's particular quality in relation to the rest of the animals is not reason, since lower animals also possess reason and consciousness to a lower degree. If we wish to establish the specific characteristic of the human being which no animal possesses, we should look for it not in reason, but in something else.

Many philosophers today put forward the idea that the differences between humanity and the animals is that we have the *urge to create our own world*. Whereas an animal faces the world as it is, and uses whatever powers it has to adjust to it, the human being also has the urge and the ability to create a world—a world of culture, history etc. People create events, institutions, festivals, not simply as means to survival or welfare, but as landmarks or points of reference for their own identity. Only human beings create art, and what art does is to take what is given, such as a tree, and make it one's own, and different, in a painting. Human culture takes an event which is past and gone, and makes of it a story, a myth, a play or a piece of national history.

Animals take what is given and respond to that in whatever way meets their immediate needs, and when human beings do the same, they share their nature with the other animals. What makes human beings unique is their ability to create, alter, or go beyond

ONLY HUMAN BEINGS CREATE ART

Fran Orford

what is given. This can even lead to human destruction of what is given. This is where the distinctiveness of humanity becomes clear. No animal would go against what is given in nature. Human beings can, and in so doing they show that their specific characteristic is not rationality but something else: freedom.

Human beings are creatures: God created us out of nothing. But God made us in his own image, and gave us both the capacity and the desire for freedom. As long as we are faced with the fact that we are created, that our being is given to us, we cannot be said to be free in the absolute sense. But we constantly see how humanity desires and struggles for absolute freedom. We are caught in a contradiction. Why did God give us such an unfulfillable drive? We suggest that God's purpose has to do with the survival of creation, with humanity's call to be the 'priest of creation'.

Questions for discussion

1 In what ways have you 'created your own world'? (In your home, for example.)
2 What does the word 'freedom' mean to you? In what ways do you wish for or struggle for freedom? In what areas do you have freedom?
3 Think of examples of the ways in which human beings do not just accept things as given, but attempt to re-make them, or to give them personal meaning.

HUMANITY'S FAILURE

The Christian view of humanity speaks of the first man and woman, Adam and Eve, as having been placed in Paradise with the order to exercise dominion over creation. It is clear that they were supposed to do this in and through their *freedom* because they were presented with a decision to obey or disobey a certain commandment from God: 'Nevertheless of the tree of the knowledge of good and evil you are not to eat.'

This commandment involved the invitation to exercise freedom: there would have been no point in giving the command if Adam and Eve had no choice about what they did. This is the freedom implied by the *imago Dei*—the freedom to act as if humans were God. This is just what Adam and Eve did, and the result is described in Genesis, chapter 3. We call it The Fall.

At this point the question arises: 'Why did humanity fall by exercising the freedom that God had given us?'

We have to find ways of interpreting the Fall other than the one involving a blame on Adam and Eve for having made too free with their freedom. The thinking of the second-century bishop St Irenaeus is helpful here.

St Irenaeus took a very 'philanthropic', very compassionate view of the fall of Adam. He thought of him as a child placed in Paradise in order to grow into adulthood by exercising his freedom. But he was deceived and did the wrong thing. What does this mean? It means that it was not a question of going too far in freedom. Adam and Eve had absolute freedom, with no limits, but they applied it in the wrong way. That is very different from saying that they should never have tried to be free, that they should have limited their freedom because they were created beings, not gods. For if they had limited their freedom this way, they would have lost the drive to absolute freedom. According to the view of St Irenaeus we can still have the drive to freedom, but we must learn to use it in the right way.

Question for discussion

How do you understand the story of the fall of Adam and Eve (Genesis, chapter 3)? What was wrong with what they did? What should they have done? What are the consequences for humanity? What are the consequences for the rest of creation?

HUMANITY, THE HOPE OF ALL CREATION

Humanity was given the drive to absolute freedom, the *imago Dei*, not for ourselves but for creation. What does this mean?

We have already discussed the idea that creation does not possess any natural means of survival. This means that if left to itself, it would die. The only way to avoid this would be communion with the eternal God. This, however, could only happen if the created world could somehow move beyond its own limitations. We need to find a way this can happen. This is why humanity was created in the image of God. Because humanity forms an organic part of the material world, being the highest point in its evolution, we are able to carry with us the whole creation towards a relationship with God.

HUMANITY'S PRIESTHOOD

Christians understand the words 'priest' and 'priesthood' in many different ways. I am using it here to refer to the relationship I have just described, where humanity acts as the representative of creation, offering ourselves and the whole created world back to God, just as the priest in the Eucharist offers the bread and wine to God on behalf of all the people. The priest has authority and a special relationship with God, but is expected to use those gifts for the benefit of all. How does this work in relation to the natural world?

We have already referred to humanity's tendency to create a new world. This tendency is what makes us different from the animals, and, in this sense, it is an essential expression of the image of God in us. This means that humans wish to pass everything that exists through their own hands and make it their own. This can result in one of the following possibilities:

1. The utilitarian way

Making it 'their own' may mean that people can use creation for their own benefit, so that creation is not lifted to the level of the human, but subjected to it.

There are two main implications of this:

(i) We would see ourselves as the most important thing in the universe, with everything else existing only in relation to ourselves—we would see ourselves as God.

(ii) We would cut ourselves off from nature as if we did not belong to it ourselves. We would understand the world as humanity's *possession*. We would use science and technology to discover ways to get the biggest possible profit from creation for our own purposes. A theology based on the belief that the essence of humanity lies in our intellect would share responsibility with science and technology for the ecological problem.

2. The personal approach

In this approach we would still use creation as a source of the basic elements necessary for life, such as food, clothing, shelter etc. But to all this we would give a dimension which we could call *personal*—that is, that humanity can only be understood in terms of our relationship with someone or something else. This could be God and/or creation. In this case, the way people eat or dress or build their houses would involve a close relationship with their environment. Humans and nature would be provided for in a community involving both of them. A personal approach to creation would lift the material world to the level of humanity's existence. The created world would in this way be freed from its own limitations. By being placed in human hands, it would itself acquire a personal dimension: it would be humanized.

The personal approach also makes every being unique and irreplaceable, not just a number in statistics. If each human being acts as a person, in a personal relationship with creation, then we not only lift creation up to the level of the human, but can also see creation as a totality: not just a collection of unrelated things, which are good, bad or indifferent according to what they do for us, but as a community, or a body, in which each part is related to the others and has its own role to play. Creation is thus able to fulfil the unity which, as natural science observes today, is inherent in its very structure.

Questions for discussion

1 In what ways do people use the created world as a set of things that are there to serve them?

2 What do you understand by the idea of living in a relationship with the rest of creation?

3 What difference does believing you are part of the same community
   make to the way one treats the world?

This way of looking at things throws a different light on the story
of the fall of Adam and Eve. It was not that creation became liable
to death because of the Fall. From the very beginning creation was
liable to die, to have an end, simply because it had a beginning.
Creation was awaiting the arrival of humanity to overcome this
predicament. The disaster of the Fall was that because of it, humanity had failed to overcome death.

If we believe Adam and Eve's fall to consist in their making
themselves the ultimate point of reference in creation, we can see
why their fall left the whole of creation under threat of death. If
creation had been lifted towards God by human hands, then
all creation could have overcome death. But by being subjected
to a created being—a human being—who assumed the place of
God, creation had no possibility of overcoming its limitations,
including the limitation of mortality. This could have been
avoided if humanity had acted as the Priest of Creation—the one
who lifts creation to God on behalf of the whole of creation.

Acting truly as a person means building relationships and
regarding the 'other', both humans and nature, as unique and
worthy of eternal existence. When a human being acts truly as a
person, then creation is treated not as a collection of mere physical
objects that are either dead or destined for death, but as something
that God meant not only to survive, but also to be fulfilled. This
fulfilling of the physical world comes about by passing through
human hands, in the way we looked at earlier. This way of acting
as a person and forming relationships has two aspects, both of
which enable us as human beings to fulfil our role as the link
between God and creation. By taking the world into our hands
and building it into a community of relationships, as well as by
referring creation to God, we liberate creation from its limitations
and let it truly be. Thus, in being the Priest of Creation in this
way, a human being is also a creator: in partnership with God we
're-create' the world. In all human creative activity there is a
priestly character—an offering on behalf of all creation.

Christians believe that what Adam failed to do Christ did:

Christ became an integral part of creation, and on behalf of the whole of creation, offered his humanity to God. He established the relationship between God and creation which creation had been unable to find. We regard Christ as the embodiment of all creation and therefore as the perfect Human Being and the saviour of the world. We regard him, because of this, as the true 'image of God'. Christ's 'bringing together' of creation and God decides the final fate of the world. We, therefore, believe that in the person of Christ the world possesses its Priest of Creation, the model of humanity's proper relation to the natural world.

It is this belief that underlies the celebration of the Eucharist. At the Eucharist we form a community which takes certain elements from creation—the bread and the wine—and offers them to God with the solemn declaration 'Thine own of thine own we offer unto thee'. We recognize that creation does not belong to us but to God, who is its only 'owner'. By both offering and recognizing God's ownership, we believe that creation is brought into relation with God. In this way it is not only treated with reverence, as befits what belongs to God, but it is also freed from its natural limitations and is transformed into a bearer of life. We believe that in doing this 'in Christ' we, like Christ, act as priests of creation. When we receive back the bread and the wine, after having offered them to God, we believe that because of this offering to God we can take them back and eat and drink them no longer as death but as life. Creation acquires for us in this way a sacredness which comes from the fact that human beings have used their freedom, their personhood, and have drawn creation into a relationship with God.

What this means is that as human beings we have an awesome responsibility for the survival of God's creation.

All of this is a belief and practice which cannot be imposed on anyone else. Because it is centred on a sacrament, a ritual, it may easily be mistaken for ritualism—nothing more than empty words and actions, with no relevance to the real world. Nevertheless we believe that all of this involves an ethos that the world, the very real, physical world, needs badly in our time. Not an ethic, but an *ethos*. Not a programme, but an attitude and a mentality. Not a legislation, but a culture.

Questions for discussion

1 There are several different ideas which may be involved in the celebration of the Eucharist (Holy Communion). What ideas are important to you?

2 The preceding section is suggesting that the celebration of the Eucharist is important for the survival of the natural world. How can this be so? Does the way a church ritual is celebrated have relevance for a global problem?

3 What do you see as being the difference between an ethic and an ethos?

4 What ethic or ethos would you put forward for the way we treat the natural world?

AN ETHOS FOR THE FUTURE

It seems that the ecological crisis is a crisis of culture. It is a crisis that has to do with the loss of the *sacrality* of nature in our culture. Sacrality means that nature is not sacred in itself, but when it is brought into association with what is holy, with God, it becomes something that is worthy of reverent treatment. If this is lost, then nature is liable to be destroyed. I can see only two ways of over-coming this. One would be the way of paganism. Paganism regards the world as sacred because it is permeated by divine presence; it therefore respects the world (to the point of worship-ping it explicitly or implicitly) and does not do damage to it. But equally, paganism never worries about the fate of the world: there is a belief in its eternity. Paganism is also unaware of any need for nature to be transformed or to rise above its limitations: the world is good as it stands and possesses in its nature all that is necessary for its survival.

The other way is what we have tried to describe here as being the Christian way. Christianity regards the world as sacred because it stands in relationship with God; thus a Christian respects it (without worshipping it, since it has no divine presence in its nature), but regards the human being as the only possible link between God and creation, a link that can either bring nature to communion with God and thus sanctify it, or turn it ultimately towards humanity and condemn it to the state of a 'thing' whose meaning and purpose go no further than the satisfaction of our needs.

Of these two ways it is the Christian view that attaches to humanity a heavy responsibility for the fate of creation. The pagan view sees humanity as *part* of the world, whereas the Christian view, by considering the human being to be the crucial link between the world and God, sees the human as the only *person* in creation—the only created being capable of creative relationships. This means that humanity, at its best, would be so deeply respectful of the impersonal world that we would wish not simply to 'preserve' the world but to cultivate and embody it in forms of culture which will raise it to eternal survival. Unless we decide to return to paganism, this seems to be the only way to respect once again the sacrality of nature and face the ecological crisis. For it is now clear that the model of human domination over nature, such as we have it in our present-day technological ethos, will no longer do for the survival of God's creation.

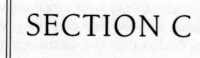

SECTION C

5 | MONASTICISM: AN ANCIENT ANSWER TO MODERN PROBLEMS

Sister Joan Chittister OSB

We have, as a people, tried every new trick we know to balance our desire for 'the good life' with its effects. We've increased our technology, multiplied our laws and expanded our educational efforts, but nothing seems to be working. Maybe it's time to try anew what worked well enough to save a civilization centuries before us so that it might save us again.

The fact is that in place of one set of values—hard work, respect for the land, simplicity, care and stewardship—our society now prefers other criteria: profit, consumption, quick returns, short-term gains and instantaneous gratification. The result, it seems, is a society that is destroying itself at the hands of its own success.

It appears that our educational system, technological society and government policies cannot provide the standards that take us beyond a thirst for things, things and more things. Where shall we go to find a model of what it takes to live in accordance with the needs of the whole earth? More than that, what will have to change in our own lives so that the world can be preserved and at the same time all life can be lived at a level of decency, beauty, health and possibility?

The answer, I think, is two-fold. First, we must begin to re-examine our theologies of creation. This has been done in Chapters 3 and 4 of this book. Then we must return to the ideals that saved Europe once and were then abandoned for short-term gain. Let us take a new look at the history and values of a way of life devised in the fifth century and dynamic to this day.

St Benedict was born about AD 480. He rejected the selfish

St Benedict with the monks. From a manuscript in the British Library.

pleasures of Rome at that time, and went to live alone in a cave 40 miles outside Rome. Others came to join him, also in search of a more simple, more Christian life, and gradually a community grew up. Later they moved to Monte Cassino, and it was there that Benedict drew up the set of rules for the life of a monastery: how to pray together, work together, learn together, and care for those in need. At the same time, Benedict's sister, St Scholastica, founded a community for women. Monastic communities spread rapidly through Europe, and almost all of them based their way of life on the rule that St Benedict wrote down. It was changed and adapted by different people at different times, but the principles of the rule have remained the same to this day.

Benedictine monasticism was a good gift for bad times. And the late fifth century was definitely a bad time for Europe. With the breakdown of the Roman Empire, the countryside was in disarray. Roads to market were prowled by thieves, the towns were unguarded and unserviced, vast properties were overrun, peasants were dispossessed, life was unsafe, unpredictable and undeveloped.

People sat and starved on untilled land or roamed and starved on unkept roadways as they searched for work and food from abandoned town to abandoned town where order was a thing of the past and markets had long been closed. The world of sophisticated cities that had been part of the legacy of Roman roads and Roman law and Roman guards and Roman administration was, for all practical purposes, over. Society had become a parade of rural villages where poor and uneducated people eked out a subsistence existence on dry and hardened land.

It was monasticism that provided the balance and security so sorely needed, and made a firm base for economic recovery. Benedictine monasticism was designed to be communal, stable and self-supporting. Unlike other religious figures of the period, monastics (monks and nuns) did not live solitary lives in desert cells or in woodland hermitages. They did not wander through the countryside begging for alms and food. Many of the saints whose names have come down to us from that period lived very austere lives: they deliberately deprived themselves of food, or shelter, or the company of others, as part of their spiritual development. Benedictines did not do this. They lived ordinary lives, but with a new stability and focus.

Benedictine monastics were formed to live a community life centred on God, in peace with all of humankind both within and outside of their own monasteries, and in harmony with nature. 'When they live by the labour of their hands', the Rule of Benedict (RB) wrote, 'as our ancestors and the apostles did, then are they truly monastics' (RB 48, 8).

Most important of all, perhaps, was the fact that the monastics themselves were tied to the land. Where their monasteries were is where they themselves would have to make a living for communities that grew rapidly and grew large. Whatever the quality of the land, they would have to till it and enhance it and harvest it and live from it. All over Europe, monastics tended the forests and put back into cultivation land that had been left barren and sterile by the barbarian multitudes that allowed it to fall into ruin. Some groups of monastics, the Cistercians, even preferred to settle in wilderness areas where they cleared the ground and turned whole uninhabitable regions into some of Europe's most fertile farmlands or reforested valleys.

Around these large, stable communities, whose land was expanded yearly both by reclamation projects and the gifts of pious benefactors, grew up villages full of people for whom the monastery became employer, school, spiritual and social centre. The monastery itself, in other words, became the local industry and social axis around which whole societies developed.

Many people gave land to the monasteries, either when they themselves entered the monastery, or as legacies when they died. This land was seldom attached geographically to the original land grant itself. Instead, the monastery fields, meadows, vineyards, forests and waters were spread across the continent. There were French monasteries with possessions in Eastern Europe, while the monastery of Fulda in Germany held land in Italy. By the year 1100, over two thousand communities were attached to one mother house in Cluny in France, living, working, functioning, letting out farms to tenants who paid in kind, and producing in like ways all across Europe. As John Henry Newman wrote, Benedictines 'were not dreaming sentimentalists, to fall in love with melancholy winds and purling rills and waterfalls and nodding groves ...'. No, these monks 'could plough and reap, ... hedge and ditch, ... drain, ... make a road, ... divert or secure the streamlet's bed', and as they approached wilderness after wilderness this way, 'the gloom of the forest departed ...' (J. H. Newman, *Historical Sketches*, II, London, 1888, 393–9).

Above all, as we shall see, each of the monasteries that lived under the Rule of Benedict operated with a vision of work and the land that marked the continent and its people for centuries.

Questions for discussion

1 Are there similarities and parallels between the fifth century, as described here, and our own time?

2 What is the greatest need of our time?

The question is, then, what did these people learn from the monasteries that enabled them to salvage a dying continent from decay and misuse that might be good news for our own time?

The answer is that Benedictine monasticism is as much a way of seeing and working and living as it is a way of praying. It is a spiritual vision that affects a person's whole style of life.

The Rule of Benedict does not deal explicitly with the management of property or the cultivation of land. What the Rule of Benedict is concerned with is the attitude that individuals take to everything in existence. As a result, this way of life has lasted for over 1,500 years and may well be as important to our own generation as it has been in times past.

Why? Because Benedict of Nursia's rule of life for monastics is not based on 'taking dominion' over the earth as some readers of the book of Genesis have emphasized. Benedict's theology of life is clearly based instead on the passage in Genesis that teaches that humanity was put in the Garden 'to cultivate it and to care for it' (Genesis 2:15).

Benedict requires five qualities of the monastic that affect the way the monastic deals with the things of the earth: praise, humility, stewardship, manual labour and community, each of them designed to enable creation to go on creating.

PRAISE

Benedictine monasticism roots a person in a community of praise. Monastics are life-positive people whose attitudes are formed as their daily services focus on the singing of the Psalms, which constantly celebrate the splendour of God in nature and the general goodness and connectedness of the cosmos. 'Sun and moon, praise God', the monastic prays weekly. 'Light and dark, wind and rain, praise God', the Psalms go on. 'Birds of the air and creatures of the sea, praise God.' No gloom and doom religion here. In monastic spirituality, in other words, everything that is, is good and to be noticed and to be honoured and to be reverenced. Nothing is expendable. Nothing is without a value of its own. Nothing is without purpose. Nothing is without beauty and quality and good.

Question for discussion

Why is praise so important for our attitude to the rest of the world?

For the person who has a monastic vision of life, then, to take from the land and not to replace it, to destroy it without reclaiming it, to have it without improving it is to violate the covenant of life. It was indeed a very monastic thing to reclaim the swamps of France and to irrigate the fields of Germany in the Middle Ages. And it is a monastic gift, in our age which destroys so much without a second thought, to recognize the value of everything, to recycle rather than to waste, to conserve energy rather than to pollute, to beautify rather than to distort an environment so that the whole world can come to praise.

HUMILITY

Benedictine humility—the idea that we each occupy a place in the universe that is unique but not compelling, wonderful but not controlling—is an antidote to excess in anything and everything. In the Benedictine view of life, monastics are to have what they need and not a single thing more: a small room, the tools of their craft, a balanced diet, plain clothes, good books. The monastic is clearly to receive whatever is necessary. On the other hand, the monastic is to hoard nothing so that others, too, can have the goods of life. 'Whenever new clothing is received, the old should be returned at once and stored in a wardrobe for the poor', the Rule reads (RB 55, 9). None of us, in other words, has an exclusive right to the fruits of creation.

It was humility and the sense of place that comes from it that led monastics of the Middle Ages to provide places of refuge for poor pilgrims. Those entering the monastery, even if they were of noble birth, were housed in common spaces and simple cells alongside uneducated peasant monks and simple labouring types. It was humility that led monastics to care for the land rather than simply to live off it.

In an age that preaches the gospel of rugged individualism and 'free-market' capitalism, monastic spirituality is a gift offered once more to a society made poor because the rich demand more riches. Benedictine humility stands with simplicity in the face of greed, conspicuous consumption and the gorging of two-thirds of the resources of the world by one-third of the people of the world, Europeans and North Americans.

STEWARDSHIP

The idea of stewardship is that one cares for something that belongs to another. As a theme, it runs right through the Rule of Benedict. The monastic is to 'care for the goods of the monastery as if they were the vessels of the altar'. The abbot is reminded that he will be required 'to give an account of his stewardship' of the monastery. 'Let him recognize', he is told, 'that his goal must be profit for the monastics, not pre-eminence for himself.' The cellarer or manager of the monastery is told to steward the resources of the community 'like a father', concerned for 'the sick, children, guests and the poor' (RB 31). Never, in other words, does monastic life or any part of it exist only for itself and its own profit. What the Benedictine monastic does and has is always for the sake of others.

In a world where control of resources, control of labour, control of profits, control of markets is the order of the day, monastic ecology calls for the cherishing of the entire planet and all of its peoples.

MANUAL LABOUR

Manual labour, the actual shaping of our private worlds, is a hallmark of Benedictine monasticism. All monastics—no matter how learned or how important—are, literally, to take life into their own hands by shovelling its mud and planting its seeds and carrying its boulders and digging its wells.

It was manual labour that made the monastic a co-creator of the universe where creation goes on creating daily. When you have

71

washed a floor and fixed a chair and painted a wall and cleared an acre and cleaned a machine, the floor and the chair and the wall and the land and the machine become important to you. You have made yourself responsible for its life.

Responsibility for life is what the modern world has most lost. In a throwaway society, nothing is seen as having life. Things have simply a temporary usefulness. As a result, we have glutted our landfills with Styrofoam cups, used once and then discarded to lie unconsumed forever while we bury the human race in its own garbage.

We are out of touch now with how long it takes to clean a polluted stream or grow a tree or dissipate a cloud of smog, and that is why we can throw bottles overboard in our lakes and waste paper by the ream and allow three people in three cars to drive to the same place day after day after day. We have indeed come a long way from the fields and the kitchens of ages past and live now in cubicles of computers and machines, whose effects on creation and life have no meaning to us whatsoever. When a man can push the button to detonate a nuclear test, it is because he has lost a sense of the monastic vision of life that comes with working to preserve it with your own hands. When a woman dumps the half-eaten casserole down a garbage disposal rather than eat left-overs, there is no sense of the monastic vision in her. When a family throws plastic cups out of a car window, it is because they have lost a sense of the value of all things that comes with the manual labour that is essential to the monastic vision of the co-creation of life.

Questions for discussion

1 How much time do you spend in the physical care of the physical world (cleaning, making, mending, cultivating etc.)? How do you feel about this time? Do you think it is too much or too little?

2 Describe something that you have made, repaired, grown etc. yourself. What are your feelings about this thing?

COMMUNITY

Finally, Benedictine monasticism is rooted in stable human community, with its variety of gifts, variety of needs, and equality amongst its members. In the monastic community of the fifth century, when slavery was considered a natural part of the human condition, the members of the monastic community lived as equals: nobles and peasants, learned and illiterate, officials and members, side by side. The only thing that distinguished the place of one monastic from another was respect for the amount of time spent in monastic life, as it formed a new kind of mentality in a society of violence and exploitation.

Here in this world where no one was to be considered the servant or the lackey or the colony of the other, everyone had equal claim on the goods of the community. Only the concept of 'enoughness' regulated the distribution of goods in the monastic community. 'Whoever needs less should thank God and not be distressed', the Rule instructs, 'but whoever needs more should feel humble because of his weakness ...' (RB 34, 3). In this society, that considered inner riches the wealth to be sought after, the notion of the accumulation of goods as a sign of character weakness was clear. It stands as a quiet comment on the patterns of conspicuous consumption and greedy capitalism that our society encourages.

It is precisely this monastic sense of praise, humility, stewardship, manual labour and community that taught Europe and made Europe fruitful and saved Western civilization. It is those things that we now, to our peril, have lost sight of.

Enlightenment for our age requires that we begin to see the planet as something with a life of its own, holy and filled with the glory of God. It is not for us to exploit or discard or use for false and short-term profit. We must begin to see the sacredness of life itself, in all its forms, for all peoples of the earth. We must begin to understand that nature is not separate from us, it is basic to us. Its fate is our fate. Its future is our future. Its life is essential to our own.

With that tradition behind them and with those concerns in mind, Benedictines today are making even more conscious efforts

73

to deal gently with their part of the globe. Benedictine monasteries do not function as a centralized government. Each monastery is autonomous and independent of other Benedictine monasteries in the world. It is impossible, therefore, to speak for all of them at once. I can, however, describe the efforts in my own community that come directly from the principles of life that we define as 'Benedictine'. Our approach to the ecological problems are four-fold: we believe that a theology of stewardship depends on our reducing, re-using, repairing and recycling the goods of the monastery so that the earth is healed and the distribution of goods throughout the world is equalized.

The four 'Rs' touch everything we do: we set out to reduce the usage of ecologically destructive products. We use no chemical additives on our grounds, not even for landscaping. All the gardening we do is organic. What we ourselves do not grow we purchase in bulk rather than in packages that are not biodegradable. We refuse to use toxic paints and unnatural materials. We wrap items for mail in old papers rather than in bubble covering. We use recycled paper in the community print shop. We cut down the amount of disposable items—cups, table covers, Styrofoam materials—that are used, even for large events.

We erected a large windmill on the property not only to provide energy for our own monastery but to determine the feasibility of using the infinitely renewable wind power in the area of the Great Lakes to save other natural resources. We compost kitchen and garden waste. We give courses on the environment at our own Camp and Conference Centre. We are founding members of the local environmental coalition that has organized many groups to salvage the part of Lake Erie where we live and to reduce the assault on the land in the locale.

The efforts are elementary but tireless as Benedictine efforts have been across time.

If we can begin to see differently and to think differently and to live differently from generations before us then we will be able to grow enough crops to feed everyone. We will not only be able to carry water across deserts but we will be able to keep it clean and clear and healthful. We will have the paper we need at reasonable cost because we will have replaced the number of trees we need to make it. We will have the national security that comes when people are not threatened either by famine or futility. We will have a world where all people are paid just wages for the work of their hands. We will have a world where being part of an 'underdeveloped' nation is a challenge, not a state of life, not a terminal disease, not an affliction without hope of cure.

Indeed, if we begin to see with monastic vision we may be able to save civilization once more:

> We will see all of life as good and refuse to dominate and diminish it.

> We will have the humility to know our place in the universe and respect, reclaim and revive the life around us.

> We will see ourselves as the stewards of the planet, not its owners, and we will pass it on to the next generation undamaged.

> We will work to shape a world full of beauty, full of possibility.

> We will build up the human community in such a way that there are no such things as 'undeveloped' peoples.

Thomas Merton wrote: 'You have to take God and creatures all together and see God in creation and creation in God and don't ever separate them. Then everything manifests God instead of hiding God or being in the way of God as an obstacle.' It is the monastic vision that calls us to see the trees and mountains of our own day as part of the glory of God and to treat them accordingly.

Questions for discussion

1 Do you consider needing things, wanting things or owning things to be a sign of character weakness? If it is, how would you set about 'strengthening' your own character?

2 How do you understand 'the sacredness of life itself'? Does a belief in the sacredness of life distract from the holiness of God?

3 The final paragraphs describe an ideal future society. Do you think this is possible? What first steps could you take towards it?

6 ST FRANCIS AND ECOLOGY

Father Peter Hooper
with Martin Palmer

Of all the saints of the Church, there is probably no better known nor more loved saint than Francis. If you want to think of a saint whose very life and words take us directly into a loving relationship with nature, St Francis immediately springs to mind. When Pope John Paul II declared a patron saint for ecology in 1979, he chose Francis. However, during his lifetime not everyone at Rome was in favour of this excited young man and at one point he was nearly expelled from the Church—in which case, we probably would only know of him as a footnote in history.

The times Francis lived in—the end of the twelfth century, beginning of the thirteenth—were times of tremendous social change. All over Europe, people were leaving the official Church and setting off on their own, or in groups, to try and come closer to Jesus. The Church had grown corrupt and wealthy. The people were suffering from poverty, overpopulation (especially in the areas of modern Belgium and the Netherlands) and the feudal lords were placing heavier and heavier burdens on the people. Power struggles between the Pope and the kings of the newly emerging countries of Europe meant that people's loyalties were split. It was not an easy time to be alive.

To many, the Church had failed. Its wealth and power meant it was part of the problem, not a solution. Its hierarchy of bishops, cardinals, archbishops and priests made both salvation and God seem remote. The Church had come to stress the divinity of Christ rather than his human nature. This made Christ a distant, severe figure. Christianity seemed to have lost its common touch, its ability to speak directly to people's lives and to their hopes and fears.

In response, new movements emerged. In the north of Europe, the Beguines and the Beghards arose. These were religious communities for women and men, where they lived in simplicity and poverty, sharing all things. While some of these stayed within the Church, others felt that the wealth of the Church meant it was no longer Christian. They attacked the Church and were eventually broken up and destroyed. In the south of France, the Albigenses and Cathars totally rejected the Church (and indeed the material world) and established their own Church. Here they practised very simple lifestyles and their priests lived in great simplicity. But they rejected many of the doctrines of the Church and so were called heretics. Great and very bloody crusades were launched against them and they were eventually brutally crushed.

Out of this upheaval came Francis. Almost single-handed he brought the Church back to the people. His teaching was centred upon the humanity of Christ and on his love and care for all his creation. Francis showed in his own life what it meant to be Christlike. His simplicity of lifestyle reminded people of Jesus. His style of preaching, telling stories about everyday things, recalled the parables of Jesus. His openness to the poor, the sick, the oppressed made him seem like a second Christ, walking again on earth. In fact, Francis is sometimes known as the Second Christ.

Francis was born in Assisi, Italy in 1182. He grew up as a wealthy young man who mixed with the young nobles and squires of the castle. He longed to be a knight and in 1202 he took part in a battle against the old enemy of Assisi, the nearby town of Perugia. The Assisi forces were routed and Francis landed up in the town prison of Perugia. Here he seems to have had time to reflect on what was really important in life. But he still determined to be a knight. Indeed he even got to the stage of setting off, fully armed, to join the Crusader army. But within a few miles of his home, he heard a voice which asked him to serve God by returning to Assisi. Reluctantly, Francis obeyed the voice.

Shortly afterwards, Francis received his call. When he was praying one day in the little church of St Damian, the great painted crucifix over the altar spoke to him and told him: 'Go, Francis, and repair my house because it is falling into ruin.' Francis, believing that the voice referred to the actual church of St Damian, set about restoring it. Then one day the crucifix spoke again: 'If you want to fulfil my will, Francis, you must scorn and abhor all that which you have desired and loved up to now.'

The effect on Francis was dramatic. He gave up his old ways and left his friends with their passion for wealth, warfare and parties. Instead he sought to love only God and to love everyone and everything he met because he saw God in them. So transformed was Francis that he gave away some of his father's wealth. This led to a furious row in front of the bishop, at the end of which Francis gave his father back all his clothes and turned, quite literally naked, to the Church for protection. From that day on Francis owned nothing but the simple habit he wore.

The poverty which Francis so willingly embraced is the hallmark of his life and of the community of brothers which he founded. He drew his model for his life of poverty from the life of Christ. When, in 1210, Francis submitted his Rule for his community to the Pope, he took as his foundation stone the poor Christ, the Christ who as the Bible says, emptied himself and took on the form of a servant. The human side of Jesus Christ's life filled Francis's mind and he saw, perhaps more clearly than anyone else before him, that Christ has become our Elder Brother, which is why we can call God Father.

In very humble circumstances, Francis and his brothers lived together, caring for the poor and the sick. In particular, the brothers travelled around the area teaching the ordinary people the love of God. In their travels, they often spent nights alone in the woods or on the mountains. This closeness to nature made a deep impression on the spirituality of the early Franciscans. To Francis himself, all nature was a sign of the love and generosity of God.

For Francis, the whole of God's creation was touched by the forgiveness and new life brought by Christ. All creation was special to God in Francis's eyes. But we must not make him into some sort of sentimental naturalist. Francis did indeed love creation, but first of all he learned to love himself by giving up his own self. Francis renounced self, self-will, self-love, self-seeking in order to give his whole self to Christ and to learn to walk in the path of Christ.

Of all his teachings, it is his ideas, beliefs and language about the rest of creation which mark Francis out from almost all other great saints and teachers of the Church. In many ways, the Church is still catching up with Francis on this question, as we shall see later.

Questions for discussion

1 What stories do you know from the life of Francis? Do they focus on his relationship with other people, his relationship with God, or his relationship with the natural world?

2 We saw in Chapter 2 (pp. 31–2) that there is not much specific information in the Bible about Jesus' attitude to the natural world. Given Francis's Christ-like life and work, can we take his attitude to the natural world as being an echo of Jesus'? Do Francis's teachings on this point fill the gap left in the gospel accounts?

Francis saw all of creation, the waters, hills and the earth itself as being loved by God and as loving God. This was totally new in the teaching of the Church. It was also considered dangerous by some. The Church in Europe had had to struggle with the older religions of the people. Many of these older religions were what might be called 'nature religions'. They saw the different parts of nature as being spirits and gods and they worshipped them. The Church, in order to preach that there was only one God, who was greater than all creation, had suppressed and destroyed these older religions. This had meant that the Church had sounded as if it did not value the different aspects of nature, the birds, animals, rivers and hills, trees, forest and clouds. God was above all this, taught the Church. But Francis turned this on its head. Having no fear of nature, he praised nature and saw nature praising God, its creator. For him, God was not the stern Father of so much of the Church's teaching. God was the creator Mother who gave birth to all and who cared, lovingly, for all. So when he called the Earth, Mother Earth, he saw her as being a part of God. When he spoke of the birds and animals as being his brothers and sisters, he meant that they were part of the same great family under God.

In his travels, Francis showed this feeling of family for the whole of nature by his attitude to such creatures as the birds. One famous story tells of his return, exhausted, from an expedition to Egypt. He had gone hoping to preach to the Crusaders and then to the Muslims. He had failed and had been disillusioned by what he saw of the Crusaders' behaviour. On landing in Italy again he was overcome by the birdsong. 'Listen', he said, 'to our brother birds who are praising the Lord; let us go among them and sing our

Francis and the birds. From a manuscript in Corpus Christi College, Cambridge.

morning service.' So saying he led the way, rejoicing in the pure simplicity of the birds' song to God. However, having joined the birds, the brothers found their own songs of praise were being drowned by the noise of the birds. So Francis spoke to the birds— as one member of a family to another. 'Brother birds, refrain from singing until we are finished with our prayers.' This the birds did until the brothers had finished their prayers.

For Francis, the coming of Jesus Christ to live amongst us was not just an event of importance for human beings. It was an event of importance for all life. Francis saw that because God loved the world, all that he had created, he gave us his greatest gift, Jesus Christ. This is not some distant God who cares nothing for the world. Here is a God who walked the earth, a God who shared the joys and sorrows, the hardships and trials of those whom he came to save. Now the world was different from what it had been before and it would never be quite the same again, for God has passed through it and left it the better for his passing.

One of the most famous stories of Francis concerns the wolf of Gubbio. The town of Gubbio was a place which Francis often liked to visit. But for some time now, its citizens had been terri- fied of leaving the city at night or alone for fear of a large ferocious wolf. The wolf attacked anyone who came by. Francis learnt of this and set off to find out what was wrong—just as a brother might go to see his brother who had gone astray. When the wolf sprang out to attack Francis, Francis spoke to him sternly and the wolf became meek and mild. It was obvious that the wolf was hungry and could not catch his normal prey. So Francis told the wolf off for acting in such an unChristian way as to attack people. He asked the wolf to show he would never attack people again. Then Francis talked to the townspeople of Gubbio. He explained the needs of the wolf and made the townspeople promise to feed the wolf and never to harm him. This, reluctantly, they did. Then Francis brought the wolf to Gubbio. Here the wolf lived happily for many years, much loved by the townspeople. When the wolf died they buried him with great ceremony and raised a statue to him. In this sort of way did Francis show the transforming power of Christ to lives both human and animal. For Francis, there was really no difference between human beings and animals—all were part of God's family. In actions such as the wolf of Gubbio, Francis saw a re-establishing of the right relationship between humans and the rest of creation, a restoration of the harmony which is the original vision of the book of Genesis.

There are two key points about Francis's love of nature. Seeing what wonders God has done for us by creating life, we should realize that we have repaid God with sin and evil. So Francis believed that we must do penance for our sins to show God that we are truly sorry for our ingratitude. This should be seen as our 'thank you' to God for all that God has done for us. Secondly, Francis saw all creatures as his brothers and sisters. This was because if Jesus could become our brother, then surely we were able to see all other creatures in the same way. Francis called those who joined him in his life of poverty, Friars—which means brothers. In using this term he showed that Christ has set the example of brotherhood, and that we humans must learn to extend this to all creation.

The greatest expression of his vision is his famous Canticle of the Creatures.

> Good Lord, most high Almighty,
> To you all praise is due,
> All glory, honour, blessing,
> Belong alone to you;
> There is no person whose lips
> Are fit to frame your name.
>
> Be praised then, my Lord God,
> In and through all your creatures,
> Especially among them,
> Through your Noble Brother Sun,
> By whom you light our day;
> In his radiant splendid beauty
> He reminds us, Lord, of you.
>
> Be praised, my Lord, through Sister Moon and all the stars;
> You have made the sky shine in their lovely light.
>
> In Brother Wind be praised, my Lord,
> And in the air,
> In clouds, in calm,
> In all the weather moods that cherish life.

Be praised, my Lord, through Sister Water,
She is most useful, humble, precious, pure.

And Brother Fire, by whom you lighten night;
How fine is he, how happy, powerful, strong.

Through our dear Mother Earth be praised, my Lord,
She feeds us, guides us, gives us plants, bright flowers,
And all her fruits.

Be praised, my Lord, through humanity
When out of love for you
We pardon one another.
When we endure
In sickness and in sorrow
Blessed are they who persevere in peace;
From you, Most High, they will receive their prize.

Be praised, my Lord, praised for our Sister Death,
From whom no one alive can hope to hide;
Wretched are they who die deep in their sin.
And blessed those Death finds doing your will.
For them there is no further death to fear.

O People! praise God and bless him,
Give him thanks
And serve him very humbly.

      (Slightly adapted from translation by Molly Riedy)

Questions for discussion

1 How difficult do you find it to think of inanimate creation—fire, water, trees etc.—praising God?

2 Take one example of inanimate creation, and make a hymn of praise that it might sing to God.

3 Is there any danger that Francis's attitude might lead one to worship nature itself?

Francis died a painful death (hence the verse about Sister Death in the Canticle) in 1226. It is said that at the moment he died, a flock of skylarks rose from the roof to accompany the soul of this great saint to heaven. He died in poverty, clad only in coarse cloth and in humble surroundings. He left an Order which then had to try and live without his shining example.

So what happened to the powerful vision of Francis, of our family relationship with all creation? If truth be told, it was almost immediately forgotten, or romanticized into a slightly soppy version of Francis who chatted to birds. The conventional understanding of nature as being below us returned and swallowed Francis's more challenging and more beautiful vision.

The Franciscan movement, established by Francis, soon relegated most of his teachings about creation to the back of its mind. Only a few Franciscans realized that Francis had opened up the possibility of seeing and relating to nature in a new way. The two best-known such people were Duns Scotus and Roger Bacon. They both lived in the thirteenth century and many have seen them as having laid the foundation stones of later Western scientific discoveries in their approach to learning and to nature. Essentially, these two men took Francis's vision of God's love being visible in all of creation, and began to look anew at creation. They looked, as a lover might look upon something which belonged to his beloved. They studied nature because it was of God. From their studies come the earliest attempts at understanding how nature worked—the basis of modern science. Yet these two men did this from love, not from some cool rational desire to take nature apart. It is a cruel irony that the reductionist approach to nature, which is more interested in the parts than in the way all parts interrelate, should have sprung from the work of these two men. Nothing could be further away from the holistic vision and understanding of St Francis.

Sadly, with the exception of these two men, Bacon and Duns Scotus, the rest of the Franciscan Orders ignored this aspect of Francis's teachings until very recently.

In the mid-1960s, after the great Second Vatican Council of the Roman Catholic Church, all the religious Orders were told to go back to their original founders and their teachings. As the Franciscans did this, they rediscovered the teachings of Francis about nature. In the years since then, Franciscans have been in the forefront of attempts to care for nature. The sad reflection is that if they had not lost this vision for so long, perhaps we would not be in quite such a terrible state today. However that may be, we can now thank God that Francis's vision has been rediscovered— maybe just in time.

Questions for discussion

1 Do you think that science and Francis's ideas can work together?

2 In what ways might scientific study be different today if Francis's ideas had remained well-known and popular?

3 What problems can you see in living life according to Francis's vision?

7 | THE PROTESTANT TRADITION

Martin Palmer

In 1517 the German monk Martin Luther nailed to the church door at Wittenberg his famous protest about corruption in the Church. In doing so he unleashed across Europe, especially northern Europe, a mass rebellion against the Roman Catholic Church. Within a few decades, it had lost control of England, Scotland, parts of Germany, Switzerland and parts of France. Sweden, Norway, Denmark and Finland had also left the Catholic Church. The Netherlands were in revolt against their Spanish overlords and Protestantism was the cause under which they gathered to resist. Across Europe, old ideas were challenged; old authorities were overthrown and new ways of thinking and behaving began to emerge.

The revolution in Church and society which Martin Luther launched changed not just the way millions thought about God and about salvation, but how they thought about nature as well.

At the heart of the Protestant understanding of life lie two key ideas which have affected how we view our relationship with nature. The first idea is that of personal salvation. Protestants believe that there is no need for a priest to act as a mediator, or bridge, between God and each person. God can be found in his Word, the Bible. Through encountering God and through a personal dedication to God through Jesus Christ, each person can be assured of salvation. The Church then is the collection of believers who are saved and who gather together to worship and study, rather than the vehicle which brings salvation itself. This affects how nature is seen because people begin to see themselves only in relation to God and not in community with the rest of his creation.

86

The second idea was that God blessed his chosen or saved ones and that a sign of this blessing was material well-being. I have a nineteenth-century book at home which shows how a down-and-out became a prosperous citizen through conversion. In the USA today, tele-evangelists, building upon, if severely corrupting, the Protestant tradition, claim that if you give yourself to God and send a donation to the tele-evangelist, then you can expect financial benefits and success in your career as a result.

Question for discussion

Do you believe that wealth and prosperity are a sign of God's blessing? What might be the problems with this view? What is the evidence for and against?

In order for people to read God's Word, the Bible, it had to be available in the languages which people actually spoke. Throughout the Middle Ages, the Roman Catholic Church had discouraged the translation of the Bible into these languages. By keeping it in Latin, the Church effectively cut people off from the Bible. Not many could read it; few could understand it when read. With the Protestant Reformation, as well as the invention of printing, suddenly millions of people were able to read the Bible. This they did with great enthusiasm. It is perhaps hard for us to appreciate how influential the Bible was in many societies. In many homes it was the only book. It was read and explored as a handbook to life both in this world and in the hereafter.

It has been said that the text in Genesis about having dominion has contributed to the situation in which people feel entitled to exploit nature. This is only partly true. Before the Reformation only clergy would have been able to read Genesis, so the people in general knew only what was passed on in preaching and teaching. Moreover, the Protestant emphasis on the centrality of the Bible was very different from the attitude of all other Churches. For them, the role of tradition, of the teachings of the great Fathers of the Church, the lives of the saints and the liturgies of the Church were as important as the Bible, sometimes even more important.

It is with the revival of belief in the key importance of the Bible and a belief that it is literally true rather than true in its broad description of life, that one can claim these texts began to have real influence amongst ordinary people, amongst the traders and business people, amongst the thinkers.

For many Protestants, the Bible is a guidebook to everyday life. In it is all the wisdom necessary for living as a Christian. Naturally, the early chapters of Genesis were especially studied, for here were answers to some of the greatest questions ever asked: Why are we here? Who made all this? How? We all want answers to these questions, for without them, we have no sense of where we belong. In the older Churches, the people found these answers in paintings, folk legends, sermons, myths and legends from other times (Greek and Roman especially) as well as from the Bible. In the Protestant world-view, the Bible held the highest authority— as it still does for many such Christians today. Therefore, we can say that with this new attitude emerging in the sixteenth century, people did begin to see those fateful verses in Genesis as some sort of Charter for Human Use of Nature.

Adam naming the creatures. From the *British Family Bible.*

By the early seventeenth century, the idea that the rest of creation had been made for our use had become part of the way of thinking for many Protestants. We find Puritan settlers in North America justifying taking the land of the native people on the grounds that they did not farm it, fell the trees or mine the minerals and therefore were not using it as God had commanded.

It is however important to say here that not all Protestants viewed nature in this way. In England, many carried forward older ideas of the value and worth of nature in God's eyes as being separate from the value and use of nature by us. The great 'metaphysical poets' of the mid- to late seventeenth century saw nature as a vehicle of God's glory and enjoyed it for its own sake. They also questioned the assumptions of their contemporaries.

> Why are we by all creatures waited on?
> Why do the prodigal elements supply
> Life and food to me, being more pure than I,
> Simple, and further from corruption?
> Why brook'st thou, ignorant horse, subjection?
> Why does thou bull, and boar so sillily
> Dissemble weakness, and by one man's stroke die,
> Whose whole kind, you might swallow and feed upon?
> Weaker I am, woe is me, and worse than you,
> You have not sinned, nor need be timorous.
> But wonder at a greater wonder, for to us
> Created nature doth these things subdue,
> But their Creator, whom sin, nor nature tied,
> For us, his creatures, and his foes, hath died.
>
> (John Donne, 1572–1631)

At the same time, the more earnest English Puritans were quite happy to follow the newer way of thinking. For instance, the Rev. Henry More wrote in 1653 that cattle showed God's goodness because by being alive, the cattle kept their meat fresh and pure until humans wished to eat it!

Questions for discussion

1 If land could be farmed to produce food for people, is it wrong to leave it uncultivated, given the amount of hunger in the world?

2 Is it possible to combine the two views, of nature being a vehicle of
   God's glory, and of nature being for humanity to use and enjoy?

The second influential idea in Protestant thought, the importance
of work and of success, is probably the most important one as far
as nature is concerned. Many people have shown the links between the Protestant ideas of the individual (standing before God,
with no priest or church to mediate) and of work, as being at the
base of the growth of an industrial, commercial, exploitative and
capitalist world-view. Indeed, capitalism has been seen and described as being a direct result of the rise of the Protestant
tradition.

What did and does this mean for nature? In earlier models of
Christianity, living in relationship with nature meant just that.
Although some other faiths are more explicit about being a part of
nature, much in traditional Christianity actually aroused a sense of
responsibility for the rest of creation. The Benedictine tradition
has already been described as an example of this.

With the rise of the idea that I as an individual can succeed, came
the idea that my success was more important than the well-being
of others—including nature. The idea emerged that not only were
we human beings apart from nature rather than the greatest part of
nature, but also, that those who were 'true Christians' were apart
from and thus above others. Also, it is about this time that one
finds phrases such as the 'conquest of nature' or the 'taming of
nature'. In part, this was due to the rise of technology and science
which made such 'control' possible or at least feasible as a goal. In
part it was an expression of our feeling that we were not only apart
from the rest of nature, but that this gave us rights over its use.
Our role was not to be the priest of creation as Metropolitan John
puts it (p. 58), but to be the victors over creation.

Here the Protestants took one idea from the older Christian
tradition and worked it further. A sense that the physical world
was not as good as the spiritual world was always present in
organized Christianity. In the Protestant faith, the fear and even at
times loathing of the physical world meant that anything physical—such as nature—was feared for its very earthiness. The spiri-

tual was given pride of place. The material, sexuality, drink, pleasure in food, dancing, physical enjoyment, were all suspect. The Puritans, who were influential in England in the seventeenth century, and who ruled England for a short time under Oliver Cromwell, were the strongest supporters of this view. For example, you may not know that the celebration of Christmas, with mince pies, decorations, food and drink etc., is still illegal, as the law passed by Oliver Cromwell in 1653 banned its celebration, describing it as a pagan festival. The same law also banned may-pole dancing, all forms of theatre, and public singing and dancing. This was one expression of the fear of the physical and material. As a result of this feeling, the world of nature was seen as savage and even downright dangerous. Not a good basis for celebrating creation in the way that so many of the Roman Catholic festivals had done and still do to this day.

One such example is Harvest festival. In many parts of the world, the Church blesses the harvest and everyone has a good time, enjoying all the fruits of God's creation and praising God for his goodness. In England, until very recently Harvest festivals celebrated how much grain or wheat we had got from the soil. We celebrated our taking from nature, not nature's bounty or God's grace in creating nature in the first place. In English Harvest festival we are observers of creation, users of it, not part of it or partakers of its joys and bounty. We do not want to be part of creation because we find the earthly world of fertility and bounty rather frighteningly physical.

Questions for discussion

1 Is the spiritual world more important than the physical world? And if so, why?

2 Look at some traditional Harvest festival hymns. What are they saying about the natural world? How often do they treat the physical world as a parable or illustration of the spiritual world?

As we saw in the chapter on St Francis, ironically, it was the Franciscan vision of the world which laid the foundations for

Christian observation and knowledge about the natural world. From this developed a rationalistic approach towards nature which led to people knowing much about a little, but little about the whole. This also fitted in well with Protestant individualism. If I can become the expert in such and such a field, then I can be of worth; I become valuable. The rise of modern science with its rationalistic and reductionist views is also traced, in part, to the rise of the Protestant tradition. In a very real way, the Protestants abandoned the idea that nature itself was of God. They could still argue that one could come to know something of God through his second Book, Nature. But what sort of God did they believe they found? At one level, it was a very distant, almost mechanical God. A God who had set the world rolling, and then had largely left it to its own devices. There was nothing 'sacred' about nature, nor indeed about this world. So much so that many Protestants came to despise the natural world and to look for its end—hardly a sympathetic vision of nature.

The most extreme expression of this came through the teachings of Calvin and his followers. John Calvin lived from 1509 to 1564. He spent his youth travelling and studying under some of the great Protestant thinkers of the time. In 1541 he went to the city of Geneva. Here he organized and directed a Protestant takeover of the city from the Catholics and the more liberal Protestants. By 1555 he had taken complete personal control of the city and imposed a Puritan dictatorship over it and every aspect of its life. Here, he believed, it would be possible to set up the perfect Christian state.

Central to Calvin's thinking was the belief in absolute predestination. Calvin believed that even before creation took place, God had decided who would be saved and who would be damned. He had predestined a few for eternal glory, the rest for eternal hellfire. Calvin believed that only those who followed his teaching would be the saved.

This attitude, in which a loving and compassionate God is replaced by a stern-faced judge who has no room for changing his mind, meant the loss of any sense that God cared for his world. All that God cared for, in Calvin's thinking, were those who he had decided would be saved. As it never occurred to Calvin that animals, plants or the very land or water might have a 'soul', thus he never saw any place for creation or indeed for the majority of the human race, in God's plan for the future.

This total dismissal of the vast majority of the world meant that Calvinism was a very successful commercial creed. If no one but the Elect—those predestined by God for glory—was of any importance, then you could use and abuse others as much as you wanted. As you were already assured of salvation, it didn't matter what you did to anyone else—your salvation was not affected. From this sprang the sort of ruthless, hardheaded business practices which helped the capitalist world become so powerful.

Furthermore, the wider Protestant idea that your success was a sign of God's blessing upon you, meant that success in the commercial and industrial world was a sign of God's blessing not just on what you did, but on how you did it.

When people attack Christianity for being the root cause of the attitude of mind which allows us to exploit nature, what they are usually referring to is this Calvinistic vision of the world.

Questions for discussion

1 Does God love Christians more than non-Christians? Is he more willing to forgive them when they do wrong?

2 The form of Calvinism discussed here is the most extreme version of this particular tradition of belief. Are there nevertheless some important points in Calvinist belief?

3 Are there elements of Calvinist beliefs, whether consciously or unconsciously, still powerfully at work in Western culture, Western science and Western Christianity?

Outside the Calvinistic tradition, there were many different attitudes, as we saw with the poem of John Donne and the views of the Rev. Henry More. In particular we must take account of movements such as the Quakers (Society of Friends) and the Mennonites. In their teachings of non-violence and of equality between peoples, they showed the love of Christ for all life. While they still exhibited much of the human-centred outlook of their age, they nevertheless saw a place for the rest of creation in the purpose of God. Listen to the words of George Fox (1624–91), one of the founders.

> What wages doth the Lord desire of you for his earth that he giveth to you teachers, and great men, and to all the sons of men, and all creatures, but that you give him the praises, and the thanks, and the glory; and not that you should spend the creatures upon your lusts, but to do good with them; you that have much, to them that have little; and so to honour God with your substance; for nothing brought you into the world, nor nothing shall you take out of the world, but leave all creatures behind you as you found them.

Quakers took a leading role in getting such bloodsports as bear-baiting and cock-fighting banned in the seventeenth century. But they were not alone. The Puritans also opposed the taking of 'delight in the cruel tormenting of a dumb creature', as Robert Bolton, writing in the mid-1620s, called it. This brings us to what was emerging from the biblical idea of dominance. On the one hand there is no doubt that many took those texts in Genesis as a divine permission to exploit the world. But others saw them as a sacred trust. They saw that animals were created for our *right* use. That is, for meat and other useful items—an attitude which most of us still have to this day. This role of animals was divinely given. Any use of animals for purposes beyond this was almost an act of blasphemy. So from an early date Puritans, as well as more liberal Protestants such as the Quakers, took a firm and principled stance against cruelty to animals. We do well to remember that the RSPCA was founded by an Anglican clergyman in the nineteenth century, who went bankrupt in order to keep the society afloat. He did this because of his religious beliefs, not despite them.

The idea of proper use of what God has given us led directly to the emergence in recent years of the idea of stewardship. A steward was a servant who looked after his master's property and had a great deal of authority over it. He made all the decisions necessary to keep the property in good order, but of course he did not own it. Christians use this image of humanity's relationship with the world. This attitude has long been there with regards to animals, as we have just seen. But the idea of extending it to the rest of creation is a new dimension, brought about by the realization of the damage we have done. It is now almost commonplace for Protestant Churches, excepting the most fundamentalist and Calvinistic ones, to support stewardship of nature. However this image raises the problem that it separates humanity from the rest of creation, setting it apart as a manager rather than a partner. Nevertheless, we should recognize that the Protestant Churches have in recent years made major attempts to alter those aspects of their traditional ways of teaching which have helped foster anti-environmental attitudes.

However, there are still some who do not feel this is appropriate. There are still those within the Protestant tradition who believe that this world is of no real significance. An example would be James Watt. An American born-again Christian of fundamentalist beliefs, he was Secretary of the Interior in President Reagan's last government. As such he was responsible for the Environment. Like a number of other fundamentalist Christians, he believed that Jesus was going to return in twenty years' time. Thus he saw no problem with licensing the destruction of forests, open-cast mining in national parks and the development of the entire coastline for housing and industry. The reason was simple. In twenty years' time Jesus would destroy this sinful wicked world; those chosen would be taken up into Heaven and saved; then Jesus would create a 'new heaven and a new earth' where the saved could live. So the whole of nature was meaningless. It is this dimension of Protestant thinking which has most deeply scarred our society's thinking. That and the idea that we have a God-given right to use the world as we wish.

It is encouraging that the Protestant Churches are now foremost in admitting this. They are engaged in a wide range of programmes which seek to restore the lost or damaged relationship with nature. Yet we still have far to go. We often remain trapped in ways of thinking and behaving which deny our expressed intention to live in relationship with nature. We inhabit a world where many of the norms and values of education, of science and technology, of business and commerce, have been shaped by the Protestant tradition, but now operate far outside the influence of any Church body. The legacy of the anti-environmental aspects of the Protestant tradition may well live on long after most of those

Protestant Churches have radically changed their attitudes. It is then that the Protestant tradition may have to take very seriously the need to speak and act strongly against the legacy of its former teachings and beliefs which so influence our structures and societies.

Questions for discussion

1 Is the view of James Watt as described above
   (a) totally wrong?
   (b) totally right?
   (c) contains some truth, but has distorted and misused it?
   What arguments would you use in support of your view?

2 If Protestant ideas have passed into secular beliefs and now exist apart from Christianity, is this a Christian problem? Should Christians be doing anything about it, and if so, what?

SECTION D

8 | TREATMENT FOR THE EARTH'S SICKNESS — THE CHURCH'S ROLE

Freda Rajotte
with Elizabeth Breuilly

There is much to be done, and the Christian Church has a large part to play in the struggle to save the earth from environmental destruction. First of all, the Church has to change itself, and show the world how to change, in several different areas: in repentance of what has been wrong; in preaching a vision of what could and should be; in giving hope and power to the hopeless and powerless; in praise and celebration of the Creator and Sustainer of the Earth; and in practical action.

REPENTANCE

The Churches have to repent of the arrogance that puts humanity at the centre of everything; of the failure to see the earth as a whole and as God's holy creation; and of division which has wounded and devalued others. It is not simply that the Church has allowed these things to happen. They are part of the whole structure of the Church. Church systems and hierarchies mean that the voices of the powerful are heard, and the voices of those who have real, lived experience of the problems are often ignored.

> O God of justice and plenty,
> whose generous earth was created
> for its own particular beauty,
> for the nourishment of its people,
> and to sing of your glory:

98

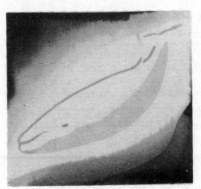

Caro Inglis

we confess that through our sinfulness
we have harvested injustice and pollution,
and not your abundance;
the land has become strange to us,
and our songs of celebration have turned harsh.

We turn to you, O God.
We renounce evil.
We seek your forgiveness.
We choose to be made whole.

(Janet Morley)

Questions for discussion

1 Make a list of actions or attitudes that have contributed to the ecological crisis. For each one, discuss who has acted wrongly and should repent.

2 As a group, try to write a suitable confession.

THE VISION

As well as turning away from the sins of the past, the Church needs to paint a clear picture of how things should be, where we are travelling to. The prophet Isaiah, besides telling his people where they had grieved God, also showed them where their aspirations towards God would lead:

99

> The wolf will live with the lamb,
> the leopard will lie down with the goat,
> the calf and the lion and the yearling together;
> and a little child will lead them. . . .
>
> They will neither harm nor destroy
> on all my holy mountain,
> for the earth will be full of the knowledge of the Lord
> as the waters cover the sea.

(Isaiah 11:6, 9)

The vision will need to have a new theology—a new understanding of God, and a new morality—a new understanding of human beings and their responsibilities.

Vision 1. A new theology

The voices from the shadows, from the oppressed, silenced and powerless, are beginning to be heard, and they are putting forward new ways of understanding the Christian faith in the light of their varied experiences. They often seem to be in conflict with each other, but all are raised in protest in God's name against the injustice and destruction that is going on, and all offer suggestions on how to change things. They are offering new theologies—and remember that theology is not a dry academic study, but a way of seeing and doing: a way of life.

For example:

- Minjung theology amongst the urban poor of East Asia re-thinks the Gospel in the context of unemployment, poverty and exploitation.

- Native American Christians are rediscovering the experience of the Cross in their own experience of the destruction of their society and culture and being torn from their lands.

- Many black theologians interpret the Gospel through the eyes of a history of slavery, colonization, segregation, apartheid and discrimination.

- In Brazil there are over 80,000 'base communities'. These are grass-roots groups of people who meet to discuss the Bible together, and in its light make practical responses to the needs of the poor. They represent the Church in action—people together, struggling for land (85 per cent of Brazil's land is owned by 5 per cent of its population), identifying with the poor (there are about 300,000 children who live on the streets in São Paulo city alone), opposing exploitation (one company funded by the World Bank recently evicted 500 families). Here the interpretation of the Bible is in the hands of people who see it as both a story and a promise of liberation from oppression.

- Feminist theology has focused on the subjection, powerlessness and poverty that women (and the earth) experience when they are under the control of men.

- There are also theologies of nature. These point to the human-centredness of traditional theology. They call for the replacement of traditional theology by a life-centred or God-centred theology, which gives reverence to all life.

These different theologies are not asking what the Church teaches about liberation, or race, or women; they are saying that the whole teaching of the Church should be rethought from the point of view of the oppressed. But which of the many oppressed people are we to listen to? Are there so many points of view clamouring for attention that we cannot take any of them seriously?

One result of such a large number of 'theologies of . . .' is to show that *all* theologies, whether past or present, express the point of view of particular sets of Christians. We are human beings, and we all have only a partial view. This makes it easier to see that traditional theology has tended to be European-centred, and to express the views of white, male and generally middle-class Christians.

One thing that all the new 'theologies of . . .' have in common, is a protest at the way people and the earth are being devalued. If people or the earth are seen to have any value at all, it is only as a means to some goal such as profit, development, modernization, or increase in national wealth. All people wish to be themselves, with their own value, their own purpose and their own fulfilment.

One of the worst things that can happen to a person is to be treated as a thing without feelings, pushed around as an object by a government, a large company, or a political party. Human beings are too precious to be treated this way—and so is the rest of creation.

These new theologies declare that both people and the land have their own value. They are not the means to some goal or value—they *are* the goal and the value: a value that is infinite in God's eyes. When we look at the unity of creation we see that people and land, the physical and the spiritual, cannot be separated. Together we compose one world.

Christians believe that it was for the life of that world that God became flesh in Christ. 'For God so loved the world that he gave his only begotten Son.'

Questions for discussion

1 (a) From the summaries given above, what 'new theology' would you like to know more of?
   (b) Which of them do you think would offer ideas that would help you or interest you?
   (c) Which of them do you see as being important in relation to the ecological crisis?

2 From your own experience of daily life, what 'theology of . . .' could you offer the world? (For example, theology of family life; of schoolwork; of the factory.) In other words, how do you see God in your own situation? Don't worry if your theology takes a negative attitude.

Vision 2. A new morality

As well as a new life-centred theology, the Church needs to find and teach a new morality for living, a new ethic. In the last 25 years an enormous amount has been written about the need to arrive at a new morality which will have at its heart a lifestyle for all which people and the earth can sustain, which can allow the healing of the sick earth. The task is a difficult one, but some important points are becoming clear.

(i) It is no longer good enough to concentrate on our own personal morality. We need to have a conscience not only as individuals, but as a nation, as a continent, as a culture. As part of whatever groups we belong to, we must look at how to eliminate injustice, poverty, famine and war. Personal morality is of course important, but for example, simply by being a member of a rich nation, many of us are part of an unjust system of trade, whether we know it or not, and whether we want it or not.

(ii) People and nature are *important in themselves*. It is a sin to treat them as merely resources or commodities. Some people have suggested establishing a code of 'ecological rights' similar to the International Declaration of Human Rights.

(iii) We have a responsibility to protect the ecology of the world for *future generations*.

(iv) Economics must recognize the *limits* of the ecological system. The earth cannot go on producing more and more. The idea of economic growth and development only means an increase in the ability to grab more of the pie: the pie itself cannot get any bigger. More land devoted to cattle for hamburger meat means less land for tropical forests. To enrich oneself at others' expense is unjust and immoral.

(v) We need a new view of our rights in relation to the land, and the land's rights in relation to us. We cannot say that the earth belongs to us; it is we who belong to the earth. The earth itself is of value in itself. The Bible affirms that 'The earth is the Lord's, and everything in it'. If it belongs to anyone or anything, under God, it is to all the creatures and living organisms that make up the biosphere. We are only temporary tenants. It is therefore wrong to buy, sell and exploit it as a commodity. In this area we have much to learn from indigenous peoples.

The idea of the land as holy needs to influence national and international decision-makers. They need to see that human activity is merely a sub-system within God's activity which sustains the life of the entire creation. In that system neither people nor the environment can be owned by anyone.

(vi) Pollution, extinction of species, destruction of forests and wildlife are crimes against the earth and against humanity. Governments and corporations must be held legally responsible for the damage they do. A series of laws and conventions are beginning to evolve to protect the environment.

(vii) It is often said that economists know the price of everything and the *value* of nothing. Why have cars and weapons of war been seen as more important than clean water and homes for children in the Third World? Isn't our very system of pricing and valuation evil in itself?

For example, in trying to find ways of cutting down on 'greenhouse' gases, some politicians are suggesting a 'tax' on pollutants. But this would in effect give governments and corporations the 'right' to pollute! This kind of thinking is very dangerous. Just as no money can give back the life of a murdered person, no money can give back the life of the earth if it is killed. We cannot put a price on things which cannot be exchanged, such as love or hope. This sort of thinking arises from the false idea that we are separate from the earth.

(viii) We have a moral obligation towards non-human creation. Christians believe in a God who declared the entire creation to be 'good' (Genesis 1:31). When God made the covenant with Noah, he made it for all the creatures (Genesis 9:12). Christ is the one 'through whom all things were made—and without whom was not anything made that was made' (John 1:3). Christ offers redemption for the whole earth (Romans 8:21). Christianity emphasizes the presence of God in and for the world.

(ix) We also have a responsibility for the suffering and harm caused to *individual animals*. Each creature is individually important to God. 'Can you not buy five sparrows for two pennies? And yet not one is forgotten in God's sight' (Luke 12:6). Those who love God will avoid causing suffering to even the least of God's creatures.

These are some of the ways of thinking which we have to change. They are important in the new ways of seeing that the Church has to learn and teach. But if we are not to be guilty of again 'dismembering the earth' in our thinking, we must turn again to the vision of the earth as a whole. The most urgent task of both science and religion is to assert the unity and sacredness of creation, and to reconsider the role of humans in it.

For example, theologians are suggesting new ways of looking at nature as the 'icon of the divine', as the 'body of God' or the 'image of God'. In some Christian traditions icons are widely used in worship. They are specially painted pictures of Jesus or the saints, which are used as a focus for prayer. The idea of an icon is that it is not divine in itself, but that it mirrors or points to God, and shows human beings something about God. God is present within creation—the divine Word that gives life to all creation.

Scientists, meanwhile, are stressing the relationships between the different systems of the world. Many view the globe as a vast and complex system made up of interacting biological, chemical and physical sub-systems. One scientist, Lewis Thomas, has studied the way the whole planet regulates its life, and he compares the earth to a vast living cell. He describes the functioning of the earth as being like a giant living creature, operating according to heart-beats or time-pulses almost too vast to comprehend:

> Viewed from the distance of the moon, the astonishing thing about the Earth, catching the breath, is that it is alive. The photographs show the dry pounded surface of the moon in the foreground, dead as an old bone. Aloft, floating free beneath the moist, gleaming membrane of bright blue sky, is the rising Earth, the only exuberant thing in this part of the cosmos. If you could look long enough, you would see the swirling of the great drifts of white cloud, covering and uncovering the half-hidden masses of land. If you had been looking for a very long geologic time, you could have seen the continents themselves in motion, drifting apart on their crustal plates, held afloat by the fire beneath. It has the organized, self-contained look of a live creature. (Lewis Thomas, *The Lives of a Cell*, Viking, New York, 1974)

Questions for discussion

1 Choose one of the points suggested for a new morality that you feel is important, and say why you have chosen it.

2 Are there any other points you would suggest for a new morality in relation to the created world?

3 If you think of the whole earth as a living being, does this alter the ways in which you treat it?

HOPE AND EMPOWERMENT

Too often, when people become aware of the immensity of the problem facing us, they give way to despair—a despair that makes them unable to do anything at all about it. What can one person possibly do that can change the course the world is set on?

It is very like the reaction of a patient in hospital who has just been told she has a serious illness. We may go through several stages: first of all saying 'It's not true—the experts have got it wrong—it's not that bad', then bargaining—'If we just reduce pollution a little and recycle more, perhaps the crisis will go away'. Finally we have to accept the fact: the crisis is here and will not go away. Once we have accepted it we can begin to do something about it.

It is not the crisis itself that holds us back from action, but the absence of hope and vision for the future. If we have hope, we are still open to possibilities. Hope gives us power to act.

So where is our hope to come from? In some cases it comes from life itself: all creatures, including people, have the will to live. But Christians have a particular source of hope in the knowledge that 'the earth is the Lord's', that creation is holy. The Bible shows that the Christian God is on the side of the poor and exploited. God shares in the poverty of the poorest, and in the brokenness and suffering of the whole creation that 'groans in travail'. In Jesus' death we see how God takes on the suffering of the world: in his resurrection we see how God overcomes the suffering of the world. The resurrection offers hope for achieving justice and peace.

> I believe that behind the mist the sun waits.
> I believe that beyond the dark night it is raining stars . . .
> They will not rob me of hope, it shall not be broken,
> it shall not be broken,
> it shall not be broken; my voice is filled to overflowing
> with the desire to sing, the desire to sing.
> I believe in reason, and not in the force of arms;
> I believe that peace will be sown throughout the earth.
> I believe in our nobility, created in the image of God,
> and with free will reaching for the skies.

Caro Inglis

> They will not rob me of hope, it shall not be broken,
> it shall not be broken.
>
> > (From a Confession of Faith from a Service for Human Rights from
> > Chile.)

For many, the Christian faith is a striving towards liberation, from political and economic oppression and exploitation, but also liberation from our slavery to material possessions and greed. Christ frees us to live with each other, for each other, with and for the whole of creation.

A world that is falling apart can only be rescued through people world-wide taking action—at grass-roots level, at parish and community level, and at national and international level. People must be given power to act—not from the top, for that takes away the power of making decisions from those who are involved in the problem. That would silence those who have the most experience and wisdom to offer—the victims themselves.

Churches cannot solve the problem of environmental destruction or poverty on their own. What they can do is stand beside and support those who are engaged in the struggle. As despair, loneliness and failure paralyse people, so faith and hope can heal and strengthen them, and restore community. Churches can help people to make moral decisions and to carry them through. They can free the Spirit that gives life to the whole world, hope to the hopeless, and strength to the weak.

Questions for discussion

1 What problems in the world are so bad that you feel nothing can be done?

2 What gives you hope for the future? (If you have no hope, say so, and say why.)

3 What people or groups do you know of who are in need of support of encouragement? How could you give that support?

PRAISE AND CELEBRATION

Let earth praise the Lord:
sea monsters and all the deeps,
fire and hail, snow and mist,
gales that obey his decree,
mountains and hills,
orchards and forests,
wild animals and farm animals,
snakes and birds,
all kings on earth and nations,
princes, all rulers in the world,
young men and girls,
old people, and children too!
Let them all praise the name of the Lord. . . .

(Psalm 148:7–13)

This new (and age-old) valuing of creation needs to be expressed in the worship, liturgies and celebration of the Church. There is much happening on this front:

— The Greek Orthodox Church has proclaimed 1 September each year to be celebrated as the Feast Day of Creation.

— The United Nations Environmental Program has established 5 June as World Environment Day. It calls upon all the world's religions to hold special Environmental worship services on the weekend closest to that day. It publishes and distributes messages from religious leaders, together with sample poems, prayers and other materials.

The multi-faith pilgrimage in celebration at Assisi.

— The WWF Network on Religion and Conservation has published the declarations from leaders of different faiths that were made at the multi-faith pilgrimage and celebration at Assisi, Italy. Since that time there have been several other celebrations and services, and new liturgies written which highlight the beauty of the earth, the creative activity of God, and the responsibility of human beings to care for the world.

> We thank you
> for you are God of all creation.
> We may call you both God and Father,
> for you hold our future in your hands,
> and all that is in this world touches your heart.
>
> Blessed are you, the source of all that exists.
> All life thirsts for you,
> because you have made us thirsty.
>
> (*Winchester liturgy*, p. 17)

> We praise you for creation
> which you made and having made, pronounced as good.
> We praise you as our source of life and light
> who pierces through the darkness we create
> and reveals to us all that vision of you
> which is in us and in all life.
>
> (Winchester liturgy, p. 18)

Reader: The Candle is lit again. Long may its light shine as we struggle to bring life again to all creation.

People of God, will you defend this tender light, this spirit which moves in and through our brother Air?

All: We will do so, through the power of the Creator Spirit.

Reader: People of God, will you go out to enlighten those around you, that the darkness of our destruction is close at hand, but here is the advent of our hope?

All: We will do so in the power of the Creator Spirit.

Reader: Go then into all God's world and share the hope that is in Christ Jesus, that all creation may know that its redeemer has come and that his light shines forth to all life.

All: In the advent of our Lord is our hope and the hope of all life. Through this we go forth to love and serve all creation. Amen.

(Advent and Ecology, p. 27)

Caro Inglis

— Around the world, at the parish level as well as from central church offices, new hymns, prayers and liturgies are appearing

110

that celebrate the earth, and praise God the Creator by reverencing all works of creation.

Response: Lord bless our land and your children who live by it.

Reader: How beautiful is the soil the Lord has made! It is rich and black and fruitful. A single seed planted in her womb will produce a hundred seeds. How beautiful is the soil the Lord has made.

Response: Lord bless our land and your children who live by it.

Reader: Who can live without soil? Can the carabao eat grass without soil? Can the wild pig survive without rooting in the soil of the forest? Even the eagle who soars above the highest mountain must return to the earth to find food.

Response: Lord bless our land and your children who live by it. . . .

(T'boli liturgy (see p. 112) quoted in Sean McDonagh, *To Care for the Earth*, Geoffrey Chapman, London, 1986, p. 163)

On a less formal level, this story comes from Burkina Faso in West Africa, where a short-term crop failure was relieved by bringing grain, not from Europe, but from nearby Ghana. This helped to support the farmers of Ghana, rather than undermining the whole local system by bringing in large amounts of grain from further afield.

As they arrived at one village home, the grandmother of the family came dancing into the open courtyard praising God for the grain, for the people who brought it, for the good crops of farmers who had grown it, and for his goodness.

(Mennonite Central Committee News Service, 1990)

Questions for discussion

1 What hymns, poems, music etc. do you know that express a sense of celebration for God's world?

2 What do you want to celebrate in your environment? Make a celebration in words, music and movement.

111

PRACTICAL ACTION

As we have seen, our beliefs, our ways of seeing, have to change. But belief without actions is a fraud. People and groups can be judged by what they are seen to be doing. So what are the churches doing about the environmental crisis? If by 'the churches' we mean all members of all Christian Churches world-wide, we are really asking, 'What is one quarter of the world's population doing about it?' Put in that context, the answer is clearly, 'Not enough!'. One quarter of the world's population ought to be able to make more of a difference. But in many parts of the world, in many different ways, a start has been made. The following are only examples:

Supporting the oppressed

— In Canada a group was formed from different Church denominations to support the land rights of Cree, Inuit, Salish and other indigenous peoples. Church people took part in blockades, marched and petitioned in solidarity with their indigenous sisters and brothers.

— In Mindanao in the southern Philippines, Sean McDonagh, an Irish Columban missionary, worked with the T'boli people, a tribal group numbering about 70,000. They have lost much of their traditional ancestral land, and they are continuing to lose it. Today they are threatened by extinction by the destruction of their environment, the rain forest, by logging companies and by those wishing to clear the forest for agriculture. They are at the bottom of the social, economic and political ladder, and the Bible tells us that they are therefore of supreme importance.

— In parts of India and Pakistan, Christian women's groups have started the 'one grain of rice' movement. Women were powerless because of poverty and because they had no control over the land, but they have formed groups and encourage each other to put aside one grain of rice a day, no matter how hungry they or their families are. These grains of rice are pooled, and as soon as they have collected a few hundred grams, the rice is sold and the money put in the women's bank. Gradually a sum of money is accumulated, enough to buy or lease land that is eroded and abandoned as unfit for growing crops. By working on this land together, the women terrace it, plant trees, and gradually bring the land back into cultivation. In this way women have become decision-makers about the land and have gained some power over their own lives.

Conservation

Many church groups work closely with local environmental and conservation projects:

— In southern India, churches became involved in the struggles of local fishing communities. They held marches and demonstrations to protest about off-shore trawling that destroys marine life and depletes fish stocks.

— In Canada, India and Australia, Church groups protested against large dam projects by guarding barricades and joining in demonstrations.

— In Austria, USA and the Soviet Union they marched against the building of nuclear power stations.

— Churches in the Pacific, in Tahiti and the Marshall Islands, in France, Australia and New Zealand have protested and petitioned against nuclear weapons testing.

Churches do not only protest at destructive projects, they are also involved in constructive projects. The A Rocha Trust was founded by a group of Christian conservationists. They run a field study centre and bird sanctuary on Portugal's Algarve coast, on the route of many migrating birds.

In Britain, the Living Churchyard Project offers project planning sheets, advice and encouragement to enable individual churches to manage their churchyards in an ecologically sound way, whether they are rural or inner city parishes.

Sustainable farming

— Some Christian religious communities, such as the Benedictines, Franciscans and Trappists, have a long tradition of the loving and careful cultivation of Church lands. In fact it was monastic orders that preserved much agricultural knowledge during the 'dark ages', spread vine-growing throughout much of Europe and beyond, and pioneered methods of plant breeding.

— In 1990 the Greek Orthodox Church established the Ormylia project. This began a project of organic farming and recycling on the monastery's lands, in an area affected by environmentally destructive practices such as extensive pesticide use. The project is to recruit a team of conservation experts to advise on organic agriculture on the monastery's land—its farm, plantations of pears, olives and almonds, and its herds of sheep, cows and chickens. They also plan to establish a local exhibition centre to spread information about conservation. Ormylia is becoming a model for surrounding farms, and for its 10,000 annual visitors.

— Several large Protestant denominations, such as the Hutterites, Amish and Mennonites, are agricultural groups who base their lifestyle upon traditional and sustainable farming practices. Hutterite communal farms are amongst the most productive examples of sustainable farming in Canada. By relying on family and community help, and using horses instead of tractors, farms are kept to what one Amish farmer describes as 'a human scale'. Farms are no bigger than what one family can take care of. The weeds in the fields are not destroyed by chemical weedkillers: some weeds are even welcomed as they help prevent topsoil from being washed away by heavy rain. Soil pests are kept at bay by crop rotation, and the farming family take delight in the number and variety of birds that make their home close to the human inhabitants of the land.

The Mennonite Centre Committee works with farmers in Africa and Asia, developing farming methods and implements which are suited to the changing conditions. For example, traditional farming methods in many parts of Africa depended on fields being left fallow for many years between crops. Now that land is more scarce, this is not possible, so Mennonite workers are helping the local farmers and the local church to investigate new methods such as the planting of trees between rows of crops, which can feed the soil without the use of chemical fertilizer.

Recycling

— In Milton Keynes, UK, an Anglican parish church established an industrial plant employing several full-time workers, to sort and package paper, glass, metals, plastics, and other products for recycling.

— The same work is done at the village built on 'Smokey Mountain', the Manila garbage dump in the Philippines.

— At a less ambitious level, parishes organize paper and bottle recycling projects, open 'good-will' stores to recycle clothing and other goods. The Salvation Army was one of the first Churches to collect used furniture and clothing and make them available to those who need them at low cost, or free.

Forest conservation

Church aid and development agencies often join forces with other groups in reforestation projects.

— The World Council of Churches recently held workshops on tropical forest conservation in Costa Rica, Indonesia and Ecuador. In these, Church organizations worked very closely with environmental groups and with local Christian communities. Equally important, they opened discussion between Church representatives and government representatives on issues of forest conservation. It may well be that this will open the way to broad government strategies with conservation in mind.

— The Vatican recently published *Tropical Forests and the Conservation of Species.*

Education

Many of the practical projects we have described include an educational side: visitors are encouraged, and the work is publicized as much as possible. This is important since people need to be aware of what can be done and is being done.

In addition, many Christian groups are taking steps towards a wider environmental education, through schools, through Sunday school material and adult study groups, and through books aimed at the general public. For example:

— Many of the aid agencies, particularly Christian Aid and CAFOD, are producing booklets, games, information packs and other educational material for schools and church groups.

— In Kenya, a series of handbooks designed to help churches in teaching the Christian faith includes specific teaching on the Christian attitude to the environment.

— In the Philippines, the Church has produced a series of books for children which teach environmental values through stories.

— Also in the Philippines, the Interlink project of the World Council of Churches has looked at some of the books used for science teaching in schools, and has come up with some very powerful criticisms of the attitudes to the environment which these books were encouraging in the students.

Policy-making

Vital though local efforts are, they are limited in their scale and effectiveness. It makes little difference how many local projects on reforestation are undertaken, if governments permit giant transnational timber companies to destroy several million hectares of tropical forest every year. Local fishing communities who practise conservation and improving fish stocks have only a tiny impact when huge fishing trawlers use radar to detect and encircle whole fish shoals, and spread drift net 'walls of death' across the oceans of the world.

To be effective at this level, Churches have to join with other activist groups, to use the media or to engage in direct political activity. This sort of activity is costly of both time and money, and can divide people at local level. And there is no guarantee of success in the end. Ultimately these problems call for international laws and controls.

Questions for discussion

1 What specific efforts to deal with the ecological crisis are you aware of? How many of them have a Christian basis?

2 What actions are you involved with, or what actions do you think you will decide to be involved with?

3 Do you think your efforts can make a difference? How could you make them more effective?

FINALLY . . .

If Churches are to play a proper part in responding to the environmental crisis they have to acknowledge the depth of the challenge. It is not nearly enough just to respond to the immediate issues: to save a few elephants here, plant a few trees there, cut down a little on carbon dioxide emissions somewhere else. The task is to analyse, criticize and transform the Western, consumerist way of thinking. That way of thinking has invaded the whole world, and will lead to environmental ruin.

Churches have always challenged people's lifestyles with ideals of simplicity and poverty. Today many Churches speak out against greed, waste and overproduction. Too much wealth has always been seen as an obstacle to the religious life. Many people have followed Jesus' words to the rich young man (Mark 10:17–22) and have given all they had to the poor, often taking vows of poverty, and have gone to serve among the poor. The Church has often made huge, even heroic efforts to 'solve' the problems of the poor. But it is just as important to challenge the system that causes the poverty.

Just as it is often easier to give 'charity' than to espouse poverty and challenge exploitation, so there is a fear that in the environmental struggle traditional Churches will only take on what fits their 'mainstream' teachings rather than allow themselves to be changed in ways that respect and reverence the whole created order.

We must bring together religious, political and environmental action. To plant trees, to protect whales and conserve marshland is to make a statement of value—that trees, whales and marshlands, in fact the whole of nature, is of great worth. The spiritual world and the material world are inseparable. The natural world is one of the ways God shows himself to us. As the German mystic Hildegard of Bingen wrote long ago:

> There is no creation
> that does not have a radiance
> be it greenness of seed
> blossom or beauty.
>
> If God had not the power thus to empower,
> the light to enlighten,
> where then would all creation be?
>
> (Gabrielle Uhlein, *Meditations with Hildegard of Bingen*, Bear & Company, Santa Fe, 1982, p. 49)

Will the Churches be able to break through the barriers of indifference and despair? Only if they accept the challenge to change themselves, their teachings and structures.

The rocks are full of the fossil remains of creatures which were unable to adapt to the changes that came over the world, and so died out. The Christian Church has to decide whether to withdraw itself from what is happening in the world and become a fossil, or to become the empowering spirit at work in and through the people committed to creation.